T0207231

Frontiers in Spray Drying

Advances in Drying Science and Technology

Series Editor
Arun S. Mujumdar
McGill University, Quebec, Canada

Industrial Heat Pump-Assisted Wood Drying
Vasile Minea

Intelligent Control in Drying
Alex Martynenko and Andreas Bück

Drying of Biomass, Biosolids, and Coal
For Efficient Energy Supply and Environmental Benefits
Shusheng Pang, Sankar Bhattacharya, Junjie Yan

Drying and Roasting of Cocoa and Coffee
Ching Lik Hii and Flavio Meira Borem

Heat and Mass Transfer in Drying of Porous Media
Peng Xu, Agus P. Sasmito, and Arun S. Mujumdar

Freeze Drying of Pharmaceutical Products
Davide Fissore, Roberto Pisano, and Antonello Barresi

Frontiers in Spray Drying
Nan Fu, Jie Xiao, Meng Wai Woo, Xiao Dong Chen

Drying in the Dairy Industry
Cécile Le Floch-Fouere, Pierre Schuck, Gaëlle Tanguy,
Luca Lanotte, Romain Jeantet

For more information about this series, please visit: www.crcpress.com/Advances-in-Drying-Science-and-Technology/book-series/CRCADVSCITEC

Frontiers in Spray Drying

Nan Fu
Jie Xiao
Meng Wai Woo
Xiao Dong Chen

CRC Press is an imprint of the
Taylor & Francis Group, an **informa** business

MATLAB ® and Simulink ® are trademarks of the MathWorks, Inc. and are used with permission. The MathWorks does not warrant the accuracy of the text or exercises in this book. This book's use or discussion of MATLAB ® and Simulink ® software or related products does not constitute endorsement or sponsorship by the MathWorks of a particular pedagogical approach or particular use of the MATLAB ® and Simulink ® software.

First edition published in 2020
by CRC Press
6000 Broken Sound Parkway NW, Suite 300, Boca Raton, FL 33487-2742

and by CRC Press
2 Park Square, Milton Park, Abingdon, Oxon, OX14 4RN

CRC Press is an imprint of Taylor & Francis Group, LLC

ISBN: 978-1-138-36473-8 (hbk)
ISBN: 978-0-367-52503-3 (pbk)
ISBN: 978-0-429-42985-9 (ebk)

Typeset in Times
by Deanta Global Publishing Services, Chennai, India

Contents

Advances in Drying Science and Technology

SERIES EDITOR: DR. ARUN S. MUJUMDAR

It is well known that the unit operation of drying is a highly energy-intensive operation encountered in diverse industrial sectors ranging from agricultural processing, ceramics, chemicals, minerals processing, pulp and paper, pharmaceuticals, coal polymer, food, forest products industries as well as waste management. Drying also determines the quality of the final dried products. The need to make drying technologies sustainable and cost effective via application of modern scientific techniques is the goal of academic as well as industrial R&D activities around the world.

Drying is a truly multi- and interdisciplinary area. Over the last four decades the scientific and technical literature on drying has seen exponential growth. The continuously rising interest in this field is also evident from the success of numerous international conferences devoted to drying science and technology.

The establishment of this new series of books entitled Advances in Drying Science and Technology is designed to provide authoritative and critical reviews and monographs focusing on current developments as well as future needs. It is expected that books in this series will be valuable to academic researchers as well as industry personnel involved in any aspect of drying and dewatering.

The series will also encompass themes and topics closely associated with drying operations, e.g., mechanical dewatering, energy savings in drying, environmental aspects, life cycle analysis, technoeconomics of drying, electrotechnologies, control and safety aspects, and so on.

ABOUT THE SERIES EDITOR

Dr. Arun S. Mujumdar is an internationally acclaimed expert in drying science and technologies. He is the Founding Chair in 1978 of the International Drying Symposium (IDS) series and Editor-in-Chief of Drying Technology: An International Journal since 1988. The 4th enhanced edition of his Handbook of Industrial Drying published by CRC Press has just appeared. He is recipient of numerous international awards including honorary doctorates from Lodz Technical University, Poland and University of Lyon, France.

Please visit www.arunmujumdar.com for further details.

Preface

The initial idea for this project was to prepare an edited book with many contributions from different authors working on different aspects of spray drying. While this would have been a platform to cover a wider range of innovative topics surrounding the spray drying process, we felt that an edited book might not provide the coherence to achieve what we envisaged for this project. At the back of our minds, we were hoping to produce a book that would coherently describe the various fundamental aspects of spray drying based on the works of a team of people who have been working collaboratively in this field for many years. At the same time, we can highlight the frontier innovations in these various aspects in our perspectives. This may give a unique flavor when reading through, which may trigger more innovations. Leaning on these ideals, we decided to embark on this project with a jointly authored book.

Looking back, the development of this book and the coherence that we feel that we have achieved is indeed reflective of how the four of us have been working together and individually on spray drying research. The book has given us the opportunity to collate and to draw upon the complementary facets of development of spray drying that we have explored over the years. It was indeed a good team effort, traversing the Pacific. Of course, this book would not be possible without the strong support and the patience from the publishing team. A special thank you to Allison and Gabrielle from CRC Press for the handling of this project and helping us with the processing of this book.

We would also like to acknowledge the Australia–China Joint Research Centre in Future Dairy Manufacturing and the National Key Research and Development Program of China (International S&T Cooperation Program, ISTCP, No. 2016YFE0101200) for providing the financial support for some of the work described in this book. Personally, Jie Xiao would like to express his appreciation for the funds from the National Natural Science Foundation of China (21978184) and the Natural Science Foundation of Jiangsu Province of China (BK20170062), both sets of funds supporting part of his effort in contributing to this book.

We would collectively like to thank Professor Arun Mujumdar for inviting us to write about spray drying in the first place. His encouragement throughout the project has been very much appreciated.

Lastly, we would like to thank our families for their patience and support for us to complete this book, especially when midnight oil had to be burned.

Nan Fu
Jie Xiao
Meng Wai Woo
Xiao Dong Chen

MATLAB ® is a registered trademark of The MathWorks, Inc. For product information, please contact:

The MathWorks, Inc.
3 Apple Hill Drive
Natick, MA 01760-2098 USA
Tel: 508 647 7000
Fax: 508-647-7001
E-mail: info@mathworks.com
Web: www.mathworks.com

Authors

Nan Fu has a Bachelor's degree in biological engineering, and two Master's degrees in biotechnology and microbiology, respectively. She obtained her PhD in chemical engineering from Monash University in Australia in 2013. She is currently an associate professor at the School of Chemical and Environmental Engineering. Her research interest is in functional powder engineering, including the fabrication of bioactive particles and food powders, functionality control, dairy science and technology, droplet drying kinetics, and particle formation mechanisms. By the end of 2019, she had published over 60 papers in international peer-reviewed journals and she is the first or corresponding author for 32 of the papers. She has presented her work in over 50 international and domestic conferences, including eight invited talks. In 2015, she was awarded the runner-up prize for the Elsevier Woman in Chemical Engineering Award (one of two awardees). In 2018, she was selected for the China-Australia Young Scientists Exchange Program (one of 16 awardees). She was also a guest editor for a special issue entitled "Functional Bioparticles" in *Powder Technology* in 2019. She is currently a youth council member of the Chinese Society of Particuology, an associate editor of the *International Journal of Food Engineering*, and a reviewer for nearly 20 journals in food and biology area.

Jie Xiao is a Jiangsu provincial chair professor, deputy head of the School of Chemical and Environmental Engineering at Soochow University, Suzhou, China. He received his BS degree in industrial automation (2001) and MS degree in control science and engineering (2004) from Zhejiang University, Hangzhou, China. In 2010, he obtained his PhD degree in chemical engineering from Wayne State University, Detroit, MI, USA. Before joining Soochow University, he was a postdoctoral research associate at Washington State University for two years. Dr. Xiao's research interests include multiscale systems science and engineering and bio-inspired chemical engineering with applied studies in spray drying systems, heat exchanger fouling and cleaning, and functional coatings. By the end of 2019, he had published around 70 refereed journal papers and delivered more than 80 talks at important chemical engineering conferences and world-leading universities. As one of the 16 awardees nationally, Dr. Xiao was selected for the prestigious 2016 China-Australia Young Scientists Exchange Program supported by the Australian Academy of Technology and Engineering and the Ministry of Science and Technology of China. He has also been recognized by the American Institute of Chemical Engineers (AIChE) as an elected senior member (2014) and has been selected by the Jiangsu Provincial Government, China, as the highest level talent for innovation (2014) and key personnel in an innovation team (2013). He was awarded the Jiangsu Provincial Chair Professorship by the Jiangsu Provincial Ministry of Education and a high-level professional talent in Suzhou by Suzhou Municipal Government, China. He is a youth council member of the Chinese Society of Particuology and an elected member of the Process Modeling and Simulation Division of the Chemical Industry and Engineering Society of China.

Meng Wai Woo is an associate professor in the Department of Chemical & Materials Engineering at the University of Auckland, New Zealand. He obtained a PhD in chemical engineering from Universiti Kebangsaan Malaysia in 2010. He carried out postdoctoral work at Monash University in Australia. He was a key player in the smart drying project funded by Australian dairy industry sectors and the project leader of an Australian Research Council-funded project on antisolvent drying to make functional particles. For the past 14 years, his main work has been on spray drying research, focusing on the computational aspect as well as the operation of the process. He has two patents pertaining to the development of unique spray dried products. Currently in the "city of sails" (Auckland), his personal interests include swimming, music, and sailing.

Xiao Dong Chen is currently a university distinguished professor and founding head of the School of Chemical and Environmental Engineering at Soochow University, China (2013–). His main professional interests include, broadly, life quality engineering, bio-inspired chemical engineering, food engineering, and particle technology. He also has a special interest in spontaneous combustion and an ongoing interest in social behavior analysis through chemical reactor engineering frameworks. Previously, he held a chair professorship at Xiamen (2010–2012), at Monash (2006–2009), and at Auckland University (2000–2005) respectively. He gained his BE in thermophysics (Tsinghua, China 1987), a PhD in chemical process engineering (Canterbury, New Zealand 1991), an MSc in applied mathematics (UNSW, Australia 2014) and was awarded a docteur honoris causa (Agrocampus Ouest 2017). He is an elected fellow of the Royal Society of New Zealand (2000–), a fellow of the Australian Academy of Technological Sciences and Engineering (2007–), and a fellow of IChemE (1999–). He is a chartered chemical engineer in the UK. By the end of 2019, he had published more than 600 journal articles and over 230 conference papers. He has led more than 60 industry projects to completion. He has delivered more than 70 invited, keynote, and plenary lectures at different institutions and conferences worldwide. He has received many awards, and some of his accolades include the Shedden Uhde Medal, 1999; ICEF-8 Young Food Engineer Award, 2000; ER Cooper Medal of the Royal Society of New Zealand, 2002; Nanqiang scholar of Xiamen University, China, 2005/2016; the inaugural Fonterra Award of Australasia, 2006; the IDS Excellence in Drying Research (one of two major awards), 2008; the IDS Founder Award for Outstanding Contribution to Drying, 2014; ICEF-12 Lifetime Achievement Award, 2015; Outstanding Award of National Education Reform and Innovation (China), 2017; *Chemical Weekly*'s Padmashri Dr GP Kane CHEMCON Distinguished Speaker (IIChE), 2017; "Next Power" chair professorship of chemical engineering at National Tsinghua University (Taiwan), 2018; ICEF-13 plenary speaker on in vitro human digestion system development, 2019; etc. He is a father of four, and his personal interests include painting, pop music, and reading.

1 Introduction

1.1 WHAT IS SPRAY DRYING?

Spray drying is a common process used to convert liquid feed (e.g., milk) into solid particles (e.g., milk powders). It is a rapid process and can be made to huge scales (from grams of powder to tens of tonnes of powders made per hour per dryer) (see Figure 1.1.1). A typical spray drying process comprises atomization of liquid droplets into a hot air flow (in the order of liquid-to-air mass ratio of about 1:10) in the drying chamber and the collection of the resulting powder. Spray drying is the single stage that turns liquid droplets to particulate format (usually at least forming a dry surface around each droplet); then, the fluidized bed(s) can be connected up (in area *D*) to further dry these particles to the desired low moisture contents and shapes/structures (see Figure 1.1.2).

The evaporation rate of the moisture (or solvent content) is determined by the difference between the vapor pressure (or concentration) of water on the droplet surface and the vapor pressure (or concentration; or more generally the humidity) of the surrounding air. Since the water content (or solvent content) of the droplet is usually quite high, the drying process starts off more like evaporation of the pure solvent droplet. Conventionally, the literature tends to say that the droplet undergoes a constant drying rate period limited by mass transfer whereby moisture evaporates and a saturated vapor film forms on the droplet surface. The droplet surface temperature approaches the wet bulb temperature of the drying air. Eventually, the wet bulb temperature will be exceeded with the build-up of solid on the surface, restricting the solvent transfer as free as for pure solvent. When the droplet surface reaches a critical (low) solvent content, the second stage of the falling-rate drying period occurs as moisture/solvent diffusion within the particle becomes the rate-limiting step. As solid concentration continues to increase, the evaporation rate decreases as more energy is required to remove the same amount of moisture/solvent and the droplet's temperature rises towards the dry bulb temperature. Increasing solid concentration increases the solvent transfer resistance, and the structure of the shell (with low permeability) influences the drying behavior. When the particle reaches the drying air temperature, the evaporation ceases at the equilibrium moisture/solvent content, which is dependent on the drying air temperature and relative humidity of the air used for drying (Kentish *et al.*, 2005; Chen and Mujumdar, 2008).

The dried particles are then collected via a cyclone separator as product after all the stages of drying operations (spray dryer plus fluidized bed dryers in sequence) that (normally for agglomeration purposes) recycles the fines back to the main drying chamber (operated from the areas *E* and *C* in Figure 1.1.1). These fines are usually encouraged by design to intersect with the semi-dried droplets (with a sticky enough surface) to form agglomerates. These agglomerates are expected to perform better in the functionality tests, in particular the wetting, dispersion, and sinking.

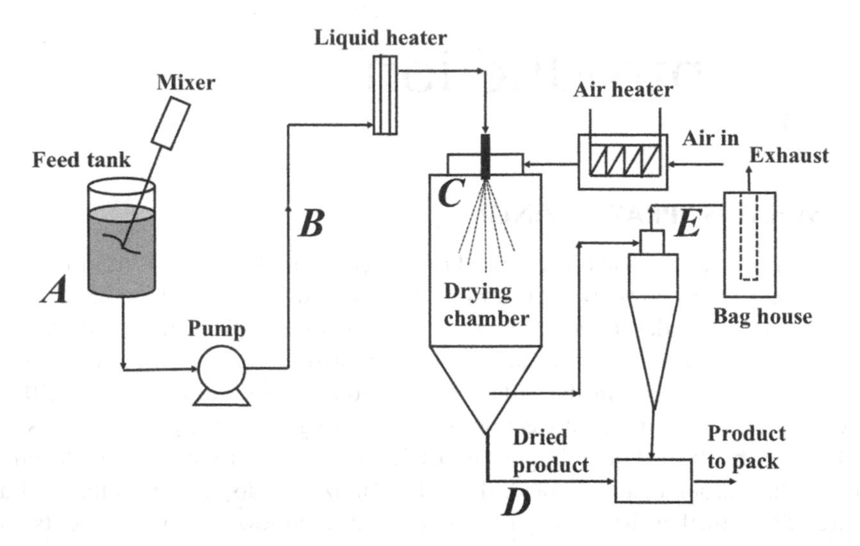

FIGURE 1.1.1 Schematic diagram showing a basic spray dryer setup with liquid preparation, atomization, drying, and product collection and classification. (Drawn by Xiao Dong Chen [XDC].)

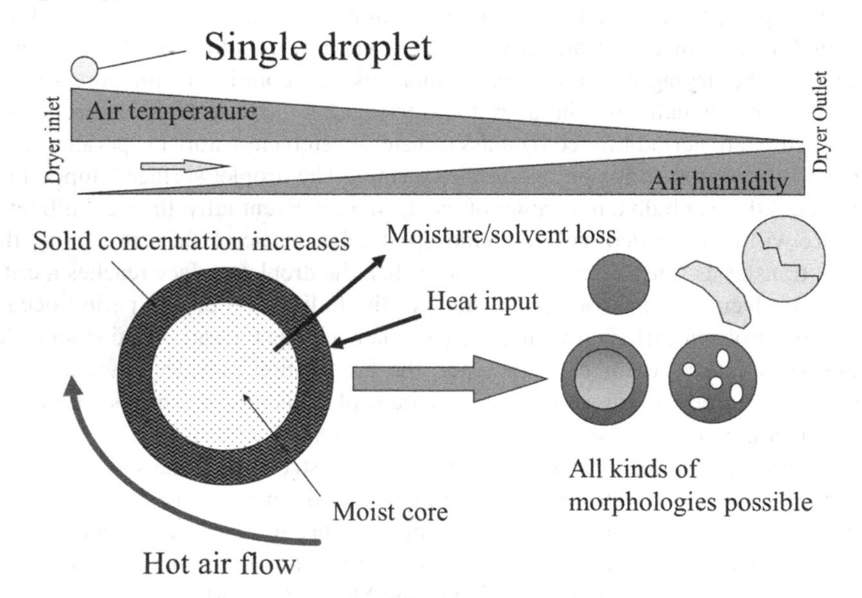

FIGURE 1.1.2 A schematic diagram showing the spray drying process to make solid particles. (Drawn by XDC.)

Furthermore, in area *A* followed by area *B*, which may be called the liquid processing section, a component separation stage (as in milk processing), pasteurization stage (for many food liquids including milk), holding and vacuum flashing stages, and concentration stage (like falling film evaporation in milk processing) can selectively or all be implemented, depending on the nature of the product. We have already seen in Figure 1.1.1 a liquid (concentrate) heating stage just before the atomization stage to assist in controlling the atomization quality. The atomization in area *C* can be carried out using a pressure nozzle (or multiple nozzles), a single rotating disk atomizer, or a two-fluid atomizer (such as an air–liquid atomizer). Nozzle atomization demands a tall-form dryer (a much taller drying chamber due to the liquid trajectory effect) and rotating disk (or wheel) atomization requires a shorter but much wider chamber for drying (due to the droplets moving down spatially and the swirling air flow).

1.2 CHRONOLOGICAL DEVELOPMENT OF THE PROCESS AND ITS MILESTONES

According to Santos *et al.* (2018) and as mentioned earlier, spray drying is a well-known particle production technology that consists of the transformation of a liquid material into dried particles. It takes advantage of a gaseous hot drying medium, usually air, and the exposure of the vast surface area of droplets to this gas medium (Masters, 1979; Cal and Sollohub, 2010). Its first observation can be dated back to 1860, when a primitive spray dryer device was patented by Samuel Percy in the United States in 1872 (Ortega-Rivas, Juliano, and Yan, 2006; Cal and Sollohub, 2010).

Ever since it was first discovered, the spray drying technique has been improved in its operational design and applications. In fact, the early spray drying devices lacked process efficiency and safety. After overcoming these issues, spray drying became a practical method for food industry purposes, ending up being used in milk powder production in the 1920s, which remains one of the most important applications until the present day. The evolution of spray drying was directly influenced by World War II, where there was an imperative need to reduce the weight and volume of food and other materials to be carried (Ortega-Rivas, Juliano, and Yan, 2006). As a result, spray drying has become an industry benchmark, namely in the production of dairy products. In the post-war period, the spray drying method continued to progress, finding application in the pharmaceutical, chemical, ceramic, and polymer industries (Vehring, Foss, and Lechuga-Ballesteros, 2007; Ortega-Rivas, Juliano, and Yan, 2006; Cal and Sollohub, 2010).

Even after more than a century of research, spray drying is still a target for study and innovation due to the increasing demand for complex particles with specific characteristics. This is considered a powerful technological process, since it brings feasibility to the production of free-flowing particles with well-defined particle size. Besides, bearing in mind the ability to use different feedstocks, as well as the technique's high productivity and broad applications, makes it a more and more effective tool for the scientific community (Ortega-Rivas, Juliano, and Yan, 2006; Patel *et al.*, 2009a,b).

Different arrangements of the gas flows and particle trajectories reduce energy losses as well as improving the ease of removing the particles from the gas medium (Huang, Mujumdar, and Chen, 2010). These arrangements can deviate considerably from the traditional design such as that shown in Figure 1.1.1. These new designs can be found at suppliers' websites, such as GEA.

People today are not only concerned about the processing aspects of the technology; they are very concerned about how good quality products can be made, especially when the products are directly related to human nutrition and health. As such, product quality has to be related way back to the processes where it might be impacted adversely in a scientific analysis. This is not an easy task and has to be carried out systematically.

1.3 MOTIVATION AND LAYOUT OF THE BOOK

Spray drying has been written about on a good number of occasions in the past. The current book has been made possible with a team of researchers who have been among the most active (and the most cited) in the area of spray drying and spray drying-related product research in recent years. The people working on this topic may well be referred to as "frontiersmen". In particular, the senior author, X.D. Chen (XDC), has been involved in this field practically and academically since 1991, when he was associated with the milk powder section at New Zealand Dairy Research Institute, working for the former incarnation of today's Fonterra Dairy Coop Ltd. He was involved in the leaderships of some of the largest dryer renovations for the best quality products at the time. The team members have collaborated closely in spray drying research at various locations and have accumulated a wealth of knowledge in the area. The kinds of emphases in the current book reflect the expertise established among the authors. Although the authors have had close working experiences in R&D with the dairy powder industries in New Zealand and Australia, their knowledge and experiences have been limited when compared with the senior technical individuals well known in the industries worldwide; hence the inevitable drawbacks of this book.

Nevertheless, it is believed that the contents of this book fit well in attempting to fill the gaps in the fundamental understanding of the related processes as well as illustrating the kind of techniques employed in recent times in process diagnoses of the spray drying industry.

The book is laid out as follows: After the short introduction in Chapter 1, Chapter 2 presents the fundamentals of single droplet drying, which provides the essential understanding and modeling approaches. This the key to understanding the whole process of spray drying. Chapter 3 looks at information about spray drying configurations, including multistage drying with spray drying as the first stage. Chapter 4 covers mass and energy balances, which cannot be avoided in any spray drying descriptions. Here, stickiness, agglomeration, and CFD (Computational Fluid Dynamics) modeling are considered. Chapter 5 uniquely addresses a special development in spray drying, i.e., spray drying of mono-disperse droplets, which has helped immensely fundamental research into spray drying, which is usually affected badly by the wide particle size distribution. Finally, variations of spray drying with hot gas-like air, such as superheated steam spray drying, crystallization during spray drying and production of bio-actives, and antisolvent vapor precipitation spray drying, are described.

2 Droplet Drying Fundamentals

2.1 THE FORMATION OF INDIVIDUAL PARTICLES DURING SPRAY DRYING

Spray drying exhibits a number of advantages as a powder production approach, including little limitation on material, no requirement of a second solvent, rapid powder production, high production capacity, continuous and automated operation, and single-step production of microparticles encapsulating bio-active substances. It thus becomes the prevailing powder production approach in a variety of industries, such as food, pharmaceutical, and chemical. In these industries, the quality of powders is of paramount importance, in addition to the efficiency of the production process. The design of a spray drying process for a specific product needs to take into account chemical product engineering and chemical process engineering. Both aspects require understanding of how powders are formed by individual droplets inside the spray drying tower.

The basic transition to form a particle in spray drying is that water evaporates from an atomized droplet and the dissolved or suspended solids are precipitated from the liquid phase and gradually solidified into a particle. In practice, the droplets may coalesce with each other, and semi-dried particles could form agglomerates of different sizes. The transition from single droplet to single particle can be rather complex, as the solids may impact the heat and mass transfer in different ways owing to the unique property of individual solids.

Figure 2.1.1 shows the particle formation processes of four dairy materials and a polymer material, recorded with the glass-filament single droplet drying (SDD) technique (Fu *et al.*, 2012). All five types of droplets contained over 85% wt of water. The initial stage of droplet drying was dominated by free water evaporation, associated with the reduction of droplet size due to the loss of water mass. As the droplet surface receded, the solids in close proximity to it could gradually enrich at the droplet surface or continue to form equilibrium with the liquid phase of reduced volume. The enrichment of solids at the surface of the droplet may alter its surface tension, as shown in Figures 2.1.1a and 2.1.1c for the skim milk and milk protein droplets. When the surface concentration of solids exceeded the critical concentration of precipitation, they could form a thin skin or a crust layer. This incipient surface layer would impact subsequent moisture removal as droplet drying progressed further. Furthermore, after drying was completed, the incipient layer would become the surface of a dried particle, which was the interface between the particle and the other medium, and hence affected a variety of important attributes of quality, such as cohesiveness, wettability, and proneness to oxidation.

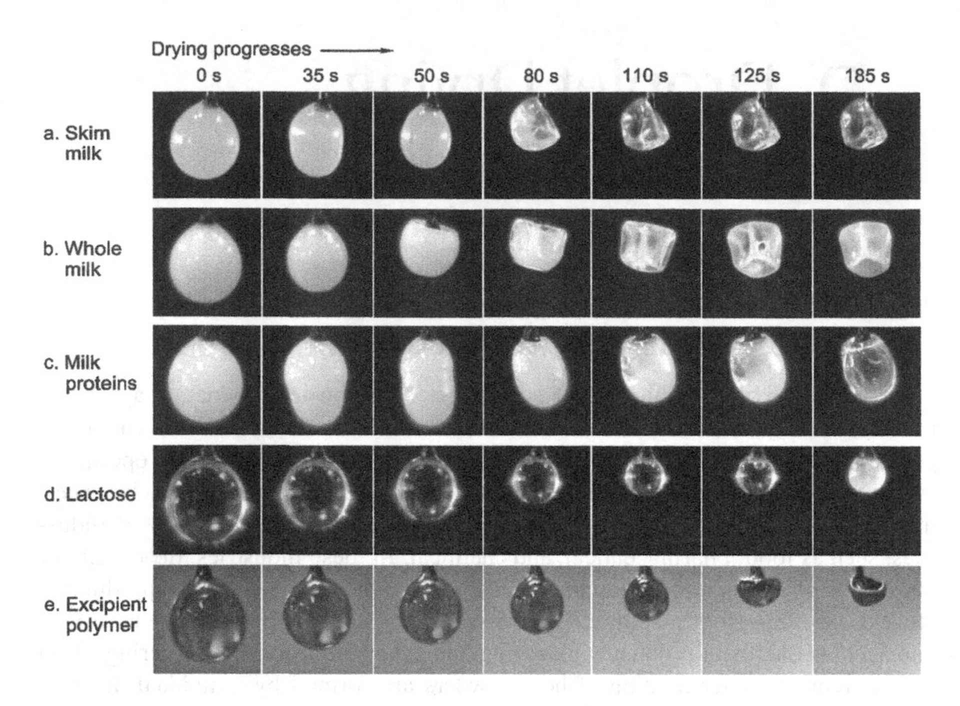

FIGURE 2.1.1 Changes in droplet morphology during drying of droplets containing different solids. (a) A skim milk droplet with initial solids content of 9% wt; (b) a whole milk droplet with initial solids content of 11.5% wt; (c) a milk protein droplet with initial solids content of 10% wt; (d) a lactose droplet with initial solids content of 10% wt; (e) a droplet containing 3% wt Eudragits® RS 30 D and 0.15% wt Rhodamine B. (Adapted from Fu, N., Chen, X.D., Chapter 14: Droplet drying, in: Liu, X.-D., Li, Z. (Eds.), *Modern Drying Technology*, 3rd ed. Chemical Industry Press, Beijing, China, 2020. With permission.)

The incipient surface layer may exhibit different properties depending on its composition. In Figures 2.1.1a, 2.1.1b, and 2.1.1c, all three types of milk solids formed an incipient layer at intermediate drying stages. The moisture contained in the interior of the droplets needed to penetrate the semi-solid surface layer to be removed, which caused further shrinkage of the droplet. The incipient surface layer formed by skim milk and whole milk droplets appeared to be soft and deformable; several spots on the layer subsided inwards, leading to dried particles with large concavities. The difference in the composition between skim milk and whole milk resulted in different colors of dried particles. Dried skim milk particles, which contained mainly lactose and milk proteins, were translucent; whereas, with the addition of milk fat, the whole milk particles became opaque. In contrast to the deformable surface layer formed by two types of milk particles, the incipient surface layer formed by milk protein droplets appeared to be rigid and resistant to deformation. The shrinkage of the incipient layer terminated after 80 s of drying, despite the continuous removal of water within it. After drying, a hollow milk protein particle was formed.

It is worthwhile noting that the incipient surface layer formed during drying of multi-component droplets could consist of a different composition from the bulk composition of the total solids. Some well-known examples include the surface coverage of milk fat (>98%) on whole milk powder (Kim *et al.*, 2002), and the over-presence of milk proteins on the surface of composite sugar–protein powder (Nijdam and Langrish, 2006). The difference between surface composition and bulk composition, also known as ingredients segregation, implies that the components in the droplets migrate during drying. The mechanisms underlying solute migration have been actively studied in recent years in order to control surface composition and to further modulate the quality of spray dried powders.

Some solids with high solubility and high diffusivity in water may slightly enrich at the droplet surface, whereas the majority of solids may continue forming equilibrium with the receding liquid phase during drying. Among the five materials shown in Figure 2.1.1, lactose was an example. The droplet shrank evenly as drying progressed, forming a spherical lactose particle (Figure 2.1.1d). The transition of transparent particle to opaque particle between 125 and 185 s of drying indicated particle crystallization of lactose. The droplet containing Eudragits® RS 30 D in Figure 2.1.1e also demonstrated even shrinkage between 0 and 110 s of drying, which was mainly due to the low solids content of 3.15% wt. When drying progressed further, a deformable incipient layer was formed, producing a buckled particle after drying.

Because of the differences in the particle formation process, spray dried microparticles can demonstrate a variety of morphologies. Some typical morphologies are shown in Figure 2.1.2. Many particles are spherical, with different degrees of wrinkles or concavities on the surface (Figures 2.1.2a–2.1.2d). As such, the surface areas of these particles are different even at similar particle sizes. During spray drying, the heat and mass transfer between atomized droplets and air occurs at the droplet surface. The different surface areas may influence the transfer phenomena and impact on droplet drying kinetics such as changes in droplet moisture content (X) and droplet temperature (T_d). In addition to the different degrees of wrinkles, spray dried particles can display various shapes, from ellipsoid to donut-like (Figures 2.1.2e–2.1.2h). A number of particles are non-spherical. Some can be oblate, with one or two flat bottoms and different degrees of wrinkles at each side (Figures 2.1.2i–2.1.2k). Some consist of thin walls and a large opening, like a pitcher (Figures 2.1.2m, 2.1.2n). Other particles may be shaped like a bowl or a plate (Figures 2.1.2l, 2.1.2o, and 2.1.2p).

The different particle shapes may influence the property and functionality of powders, and affect further processing operations. For example, powders that have larger surface area per unit mass may exhibit reduced flowability, because of the enlarged surface area for inter-particle bonding and interactions. Dense spherical particles would be preferred in tableting, as they contain less occluded air. The control of particle morphology formed by different solids, so as to precisely control the quality of spray dried powders, has received growing attention in recent years.

2.1.1 The Drying Kinetics of Droplets during Spray Drying

As different solids can demonstrate a variety of particle formation processes, the heat and mass transfer between the solid-containing droplet and drying air is expected to

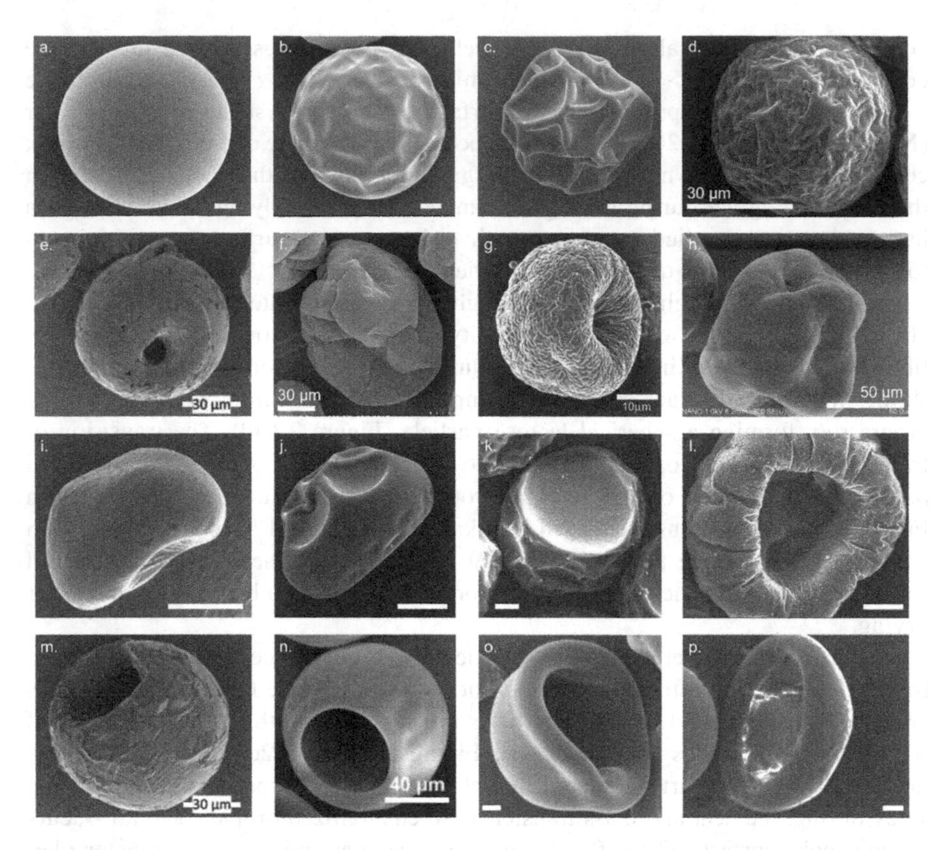

FIGURE 2.1.2 The various morphologies shown by microparticles produced by spray drying. Unlabelled scale bars represent 10 µm. (Adapted from Fu, N., Chen, X.D., Chapter 14: Droplet drying, in: Liu, X.-D., Li, Z. (Eds.), *Modern Drying Technology*, 3rd ed. Chemical Industry Press, Beijing, China, 2020. With permission.)

be different. As a result, atomized droplets can experience varying drying kinetics, even under the same drying conditions. Common kinetic parameters of a droplet include: droplet mass (m), droplet temperature (T_d), droplet moisture content (X), which can be calculated with m and the solids content of the droplet (w_s), and their derivatives, drying rate (dm/dt), rate of temperature change (dT/dt), and rate of moisture removal (dX/dt). The change in droplet surface area (A) is also of high importance, as heat and mass transfer occurs at the droplet surface. It is often evaluated with the characteristic length of the droplet drying system, namely, the diameter of the droplet (D).

Figure 2.1.3 shows the drying kinetics of a lactose droplet, a skim milk droplet, and a whole milk droplet during convective drying at 90°C. Three types of droplets had the same initial droplet size of 2 µL and similar initial solids content around 10% wt, but their drying kinetics were notably different. The temperature profiles of the three types of droplets showed a similar pattern (Figure 2.1.3a), which was characteristic in drying of dilute droplets (Rogers *et al.*, 2012). In contact with hot air,

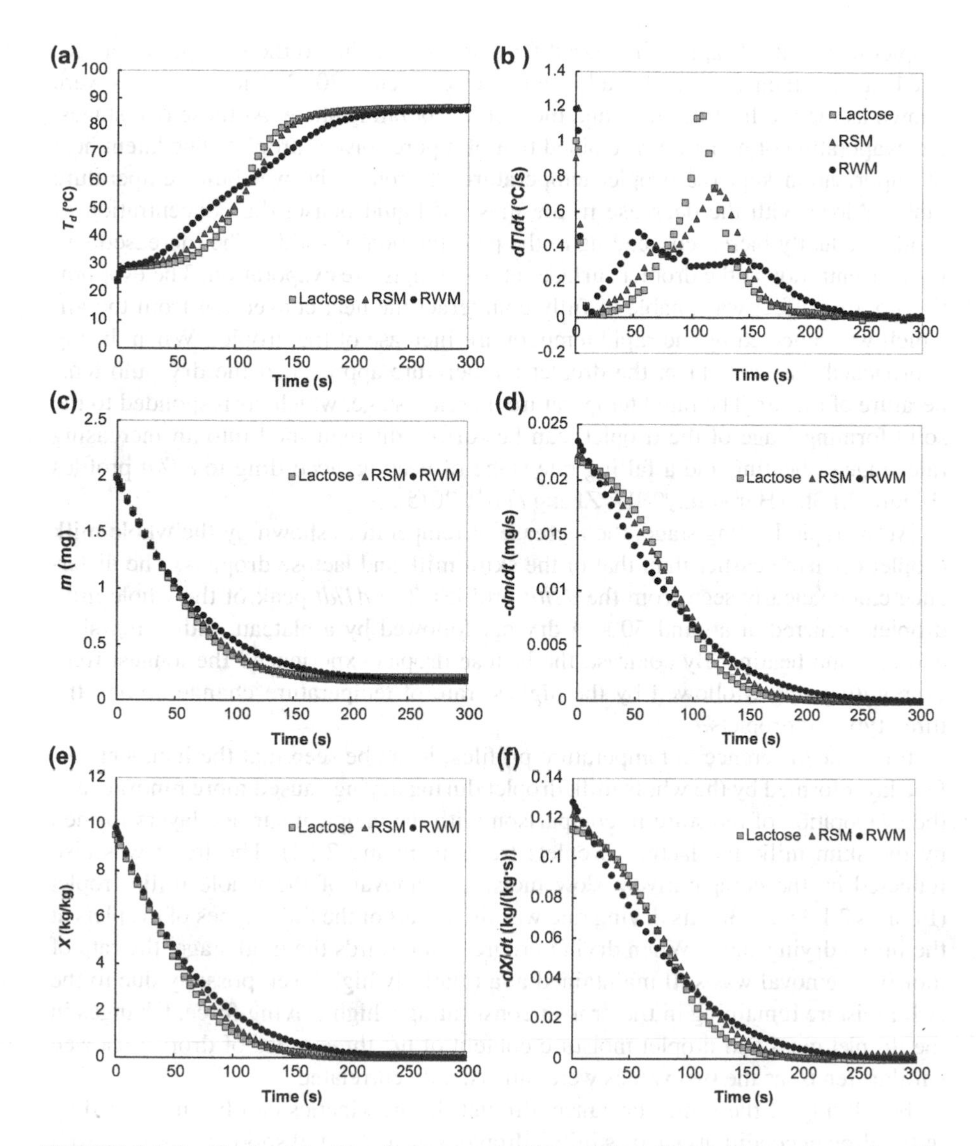

FIGURE 2.1.3 Drying kinetics of a lactose droplet, a skim milk droplet, and a whole milk droplet during convective drying at 90 °C. The initial droplet size was 2 μL, and the initial solids content was around 10% wt. The drying air conditions were velocity of 0.75 m · s^{-1} and humidity of 0.001 kg · kg^{-1}. RSM: reconstituted skim milk; RWM: reconstituted whole milk. Figures (a–f) show the profiles for droplet temperature, the rate of temperature change, droplet mass, drying rate, droplet moisture content, and the rate of moisture removal, respectively. (Adapted from Zheng, X., Fu, N., Duan, M., Woo, M.W., Selomulya, C., Chen, X.D, The mechanisms of the protective effects of reconstituted skim milk during convective droplet drying of lactic acid bacteria. *Food Research International* 76, 478–488, 2015.)

droplet temperature rapidly increased for a short time due to the receipt of convective heat, and then stabilized at a low temperature below 30°C. The two stages were known as the pre-heating stage and the free-evaporating stage. At these two stages, the evaporation of moisture resembled that of a pure solvent droplet. The latent heat of vaporization kept the droplet temperature at around the wet bulb temperature range. Along with the decrease in the mass of liquid phase, the concentration of solids gradually increased, leading to the precipitation of solids. The increased solids concentration at the droplet surface retarded moisture evaporation. The evaporative cooling effect was unable to fully counteract the heat convection from the air, which was reflected by the rapid temperature increase of the droplet. When drying approached the final stage, the droplet temperature approached the dry bulb temperature of the air. The rapid temperature increase stage, which corresponded to the solid-forming stage of the droplet, can be further distinguished into an increasing rate stage of heating and a falling rate stage of heating, according to dT/dt profiles (Figure 2.1.3b) (Har *et al.*, 2017; Zheng *et al.*, 2015).

At the rapid heating stage, the increase of temperature shown by the whole milk droplet occurred earlier than that of the skim milk and lactose droplets. The difference can be clearly seen from the dT/dt profiles. The dT/dt peak of the whole milk droplet occurred at around 50 s of drying, followed by a plateau, indicating slow and constant heating. By contrast, the lactose droplet experienced the longest free-evaporating stage, followed by the highest rate of temperature change among the three types of droplets.

From the difference in temperature profiles, it can be seen that the incipient surface layer formed by the whole milk droplet during drying caused more hindrance to the evaporation of moisture in comparison with the incipient surface layers formed by the skim milk and lactose droplets (refer to Figure 2.1.1). The trend was also reflected by the comparatively slow moisture removal of the whole milk droplet (Figures 2.1.3c, 2.1.3e). Its drying rate was the lowest of the three types of droplets at the initial drying stage. When drying progressed towards the final stage, the rate of moisture removal was still maintained at a relatively high level, possibly due to the rich moisture remaining in the droplet constituting a high driving force. Changes in the droplet mass and droplet moisture content of the three types of droplet showed similar trends, as the two values were intrinsically correlated.

For drying of the same substance, droplet drying kinetics can be influenced by initial droplet conditions such as initial droplet size and initial solids content, as well as by drying air conditions such as air temperature, air velocity, and the humidity of air. Figure 2.1.4 shows the temperature and mass profiles of 10% wt lactose droplets with three initial sizes during drying at three temperatures. The nine temperature curves and nine mass curves separated well from each other. All temperature curves showed similar patterns with the four characteristic stages, namely, pre-heating, free-evaporating, heating due to solid-forming, and the final stage. For lactose droplets of the same size, increasing the drying temperature led to a shortened free-evaporating stage, increased rate of heating, and a high final temperature of the dried particles (Figure 2.1.4a). The trend was in line with the increased drying rate as shown in Figure 2.1.4b. At the same drying temperature, droplets of smaller sizes dried faster because they contained less moisture (Figure 2.1.4a). The corresponding

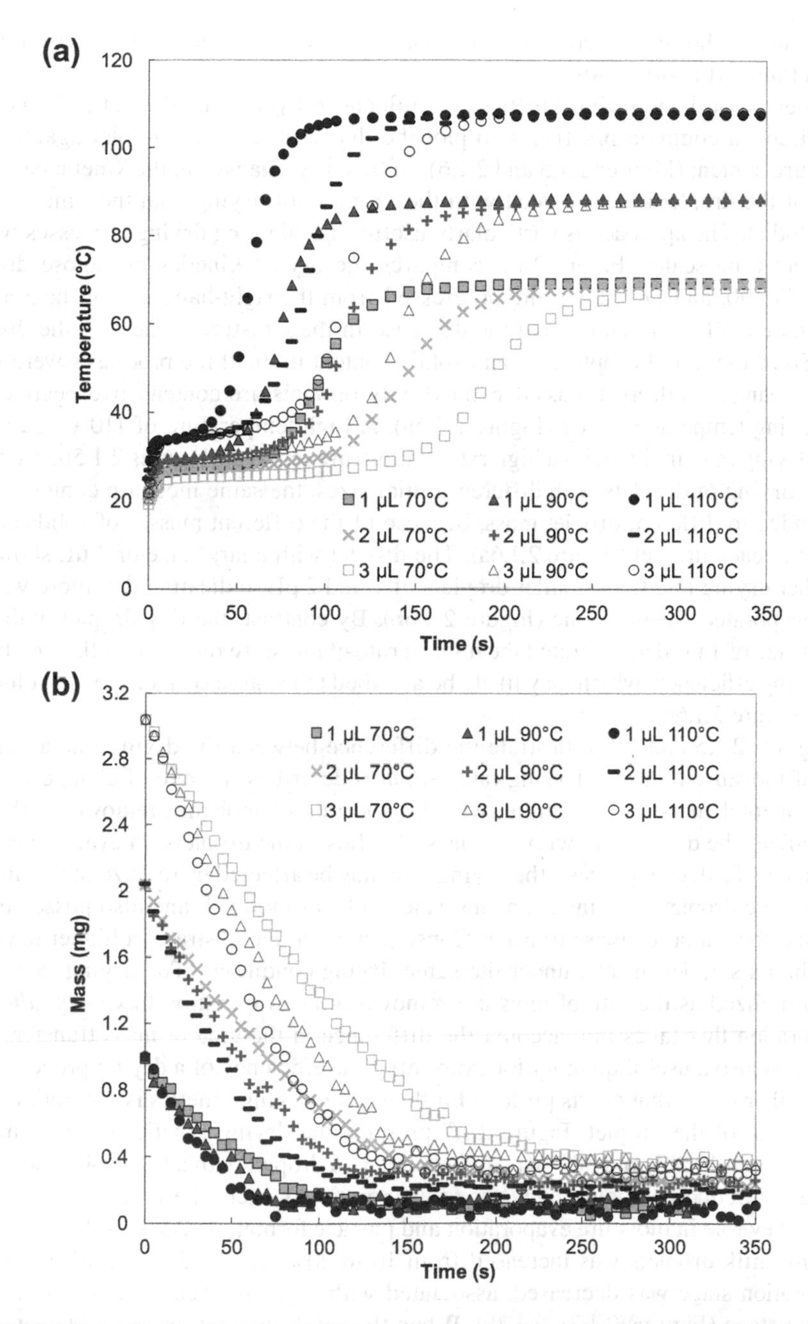

FIGURE 2.1.4 Drying kinetics of 10% wt lactose droplets with three initial droplet sizes during drying at three air temperatures. The velocity of the air was 0.75 m · s^{-1} and the humidity was 0.0001 kg · kg^{-1}. (a) Droplet temperature profiles; (b) droplet mass profiles. (Reprinted from Fu, N., Woo, M.W., Lin, S.X.Q., Zhou, Z., Chen, X.D., Reaction Engineering Approach (REA) to model the drying kinetics of droplets with different initial sizes – experiments and analyzes, *Chemical Engineering Science* 66(8), 1738–1747, 2011.)

mass curves showed different starting points due to the differences in the initial droplet mass (Figure 2.1.4b).

To examine how drying kinetics are influenced by various droplet and drying conditions, a common practice is to plot the drying rate of the droplet against its moisture content (Figures 2.1.5 and 2.1.6). In this way, changes in the kinetic parameters of the droplet are correlated with the progress of drying, and the time factor is excluded. The approach is particularly useful in evaluating drying processes with different time scales. Figure 2.1.5 compares the drying kinetics of lactose droplets at 70, 90, and 110°C. Drying progressed from the right-hand end of the *x*-axis towards the left-hand end, along the decrease in the moisture content of the droplets. Since the initial droplet size and solids content in the three processes were the same, changes in droplet mass depended only on moisture content, irrespective of the drying temperature used (Figure 2.1.5a). A high temperature of 110°C led to a high drying rate, in line with a high rate of moisture removal (Figures 2.1.5b, 2.1.5c). In the drying of droplets with different initial sizes, the same moisture content corresponded to different droplet mass, because of the different masses of solids contained in each droplet (Figure 2.1.6a). The droplet with a large size of 3 μL showed a higher drying rate than smaller droplets of 1 and 2 μL, indicating that more water was evaporated per unit time (Figure 2.1.6b). By contrast, the tiny droplet with an initial size of 1 μL demonstrated the highest rate of moisture removal, indicating better drying efficiency, which may likely be ascribed to its large surface area to volume ratio (Figure 2.1.6c).

Figures 2.1.5 and 2.1.6 illustrate the difference between the drying rate and the rate of moisture removal. Drying rate ($-dm/dt$) describes the rate of change in the mass of total moisture over time, whereas the rate of moisture removal ($-dX/dt$) normalizes the drying rate with the mass of solids in the droplets. In evaluating the efficiency of a drying process, the drying rate may be affected by the size of the droplet. A large droplet contains abundant water to be evaporated, and also possesses a large surface area for mass transfer. Consequently, it demonstrates a higher drying rate than a smaller droplet under the same drying conditions. The drying rate can be normalized as the rate of moisture removal or as evaporation flux ($-dm/(dt{\cdot}A)$). Evaporation flux takes into account the difference in the area of mass transfer. As such, it is also a useful quantity for examining the efficiency of a drying process.

Another factor that exerts profound influence on drying kinetics is the initial solids content of the droplet. Figure 2.1.7 presents the drying kinetics of skim milk droplets with 10, 20, and 50% wt solids content. A droplet with a high solids content contains a reduced amount of water to evaporate. As such, the property of solids plays a key role in moisture evaporation and particle formation. As the solids content of skim milk droplets was increased from 10 to 20% wt, the duration of the free-evaporation stage was decreased, associated with an early occurrence of the rapid heating stage (Figures 2.1.7a, 2.1.7b). When the solids content reached 50% wt, the droplet temperature rapidly increased towards the dry bulb temperature from the start of drying. There were no pre-heating and free-evaporation stages, indicating that the water content at the droplet surface was too low to balance the convective heat transfer. Milk solids constituted the main resistance to mass transfer immediately after drying was started. In line with the temperature profiles, the skim milk

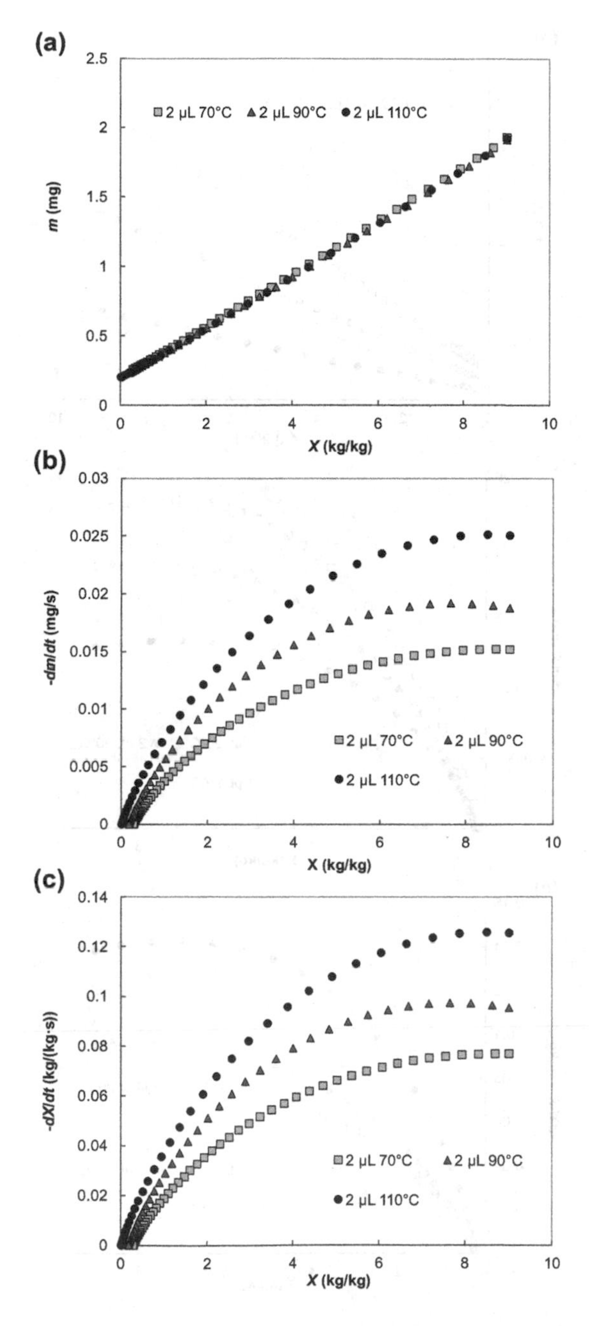

FIGURE 2.1.5 Drying kinetics of 10% wt lactose droplets at three air temperatures, plotted against the corresponding moisture content of the droplet. The initial droplet size was 2 μL. The velocity of the air was 0.75 $m \cdot s^{-1}$ and the humidity was 0.0001 $kg \cdot kg^{-1}$. (a) Droplet mass; (b) drying rate; (c) rate of moisture removal. (Figures b and c adapted from Fu, N., Chen, X.D., Chapter 14: Droplet drying, in: Liu, X.-D., Li, Z. (Eds.), *Modern Drying Technology*, 3rd ed. Chemical Industry Press, Beijing, China, 2020.)

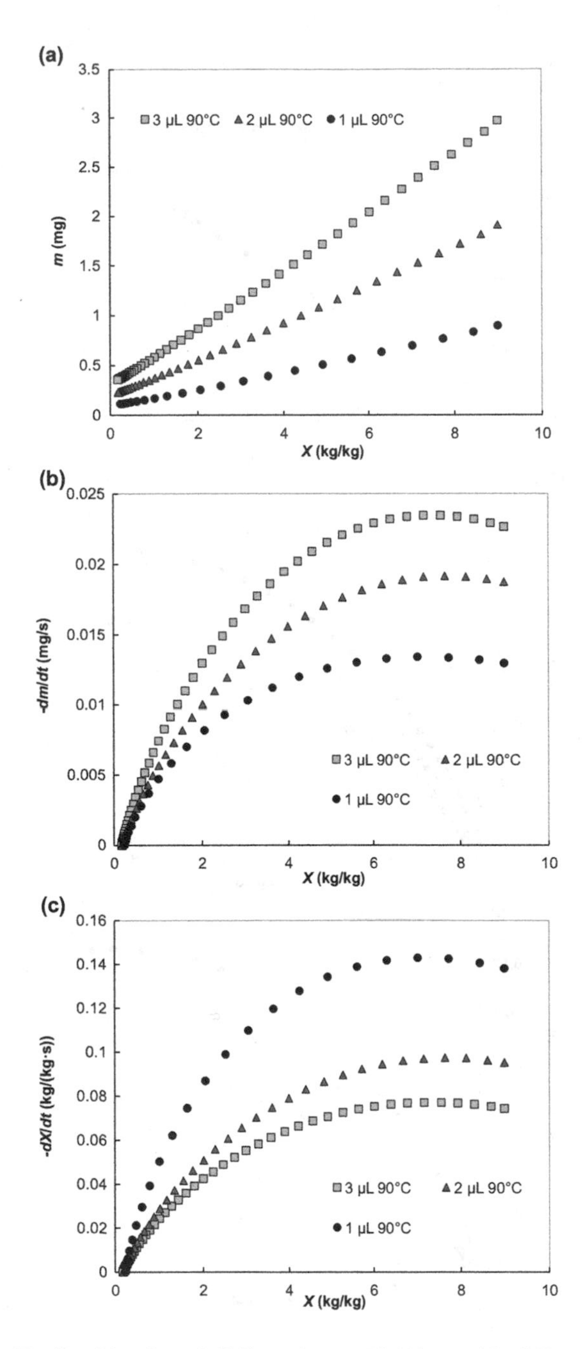

FIGURE 2.1.6 Drying kinetics of 10% wt lactose droplets with different initial sizes at 90°C plotted against the corresponding moisture content of the droplet. The velocity of the air was 0.75 $m \cdot s^{-1}$ and the humidity was 0.0001 $kg \cdot kg^{-1}$. (a) Droplet mass; (b) drying rate; (c) rate of moisture removal. (Figures b and c adapted from Fu, N., Chen, X.D., Chapter 14: Droplet drying, in: Liu, X.-D., Li, Z. (Eds.), *Modern Drying Technology*, 3rd ed. Chemical Industry Press, Beijing, China, 2020.)

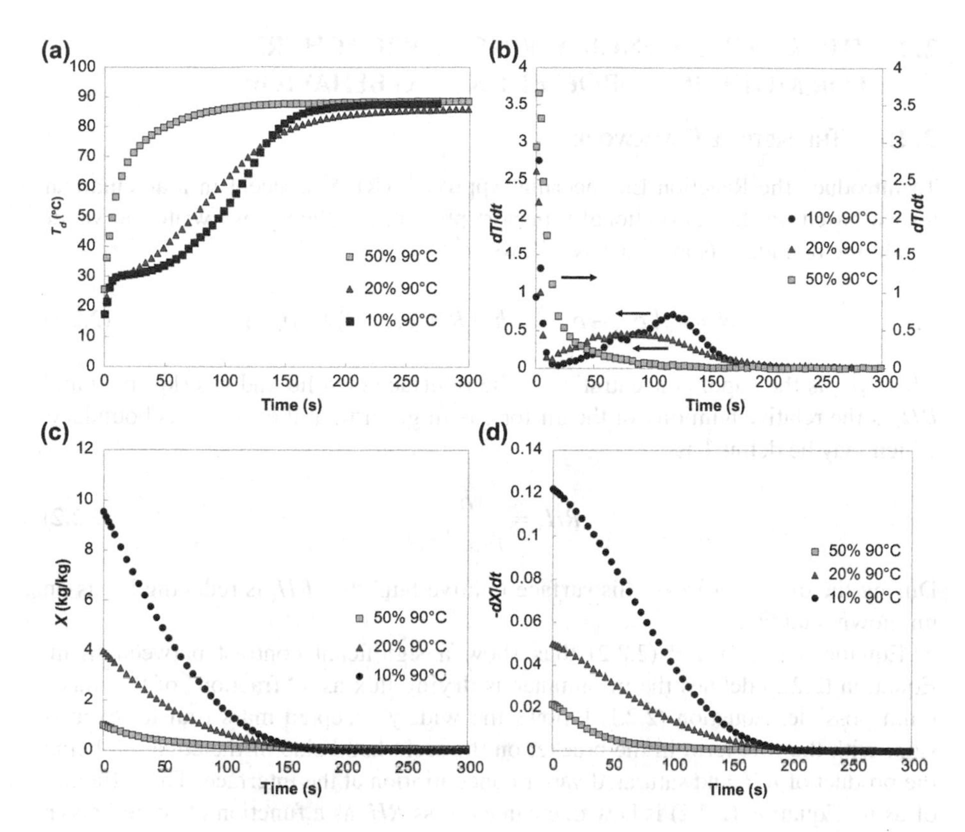

FIGURE 2.1.7 Drying kinetics of skim milk droplets with different initial solids content at 90°C. Skim milk was prepared by reconstitution. The initial droplet size of 10% and 20% wt droplets was 2 μL, and that of 50% wt droplets was 1 μL. The drying air conditions were velocity of 0.75 $m \cdot s^{-1}$ and humidity of 0.0001 $kg \cdot kg^{-1}$. Figures (a–d) show the profiles for droplet temperature, rate of temperature change, droplet moisture content, and rate of moisture removal, respectively. (Adapted from Fu, N., Huang, S., Xiao, J., Chen, X.D., Chapter Six – Producing powders containing active dry probiotics with the aid of spray drying, in: Toldrá, F. (Ed.), *Advances in Food and Nutrition Research*. Academic Press, pp. 211–262, 2018.)

droplet of 50% wt solids content also showed the lowest rate of moisture removal among the three types of droplets (Figure 2.1.7d).

From the above discussion, the drying kinetics of atomized droplets vary substantially with differences in solids material, air conditions, and droplet conditions. In spray drying practice, the atomized droplets may have vastly different sizes, and sometimes their composition can be slightly different when multi-component feed is sprayed. Moreover, inside the spray drying tower, atomized droplets may undergo different trajectories with changing environmental conditions. As a result, the billions of atomized droplets inside the dryer may show distinct drying kinetics. The difference in drying kinetics affects many quality attributes, such as crystallization and the retention of bio-active substances. To perform numerical simulation of a spray dryer, a robust mathematical model capable of describing the varying droplet drying kinetics is of paramount importance.

2.2 THE REACTION ENGINEERING APPROACH (REA) FOR MODELING DROPLET DRYING BEHAVIOR

2.2.1 Theoretical Framework

To introduce the Reaction Engineering Approach (REA) concept in mathematical terms, based on the conventional transport phenomena theory, we write the vapor flux at the boundary (solid–gas) as follows:

$$N = h_m\left(\rho_{v,s} - \rho_{v,\infty}\right) = h_m\left(RH_s\rho_{v,\text{sat}}\left(T_s\right) - \rho_{v,\infty}\right) \tag{2.2.1}$$

where $\rho_{v,s}$ is the vapor concentration at the interface of solid and gas ($\text{kg}\cdot\text{m}^{-3}$) and RH_s is the relative humidity of the air (or gas in general) at the solid–gas boundary, which may be defined as:

$$RH_s = \frac{\rho_{v,s}}{\rho_{v,\text{sat}}\left(T_s\right)} \tag{2.2.2}$$

During the drying process, this surface relative humidity RH_s is reducing but is an unknown quantity.

Equations (2.2.1) and (2.2.2) thus show a significant contrast between them. Equation (2.2.2) defines the instantaneous drying flux as a "fraction" of the maximum possible; Equation (2.2.1) follows the widely accepted mass transfer expression, with the first term in the bracket on the right-hand side of the equation being the product of RH_s and saturated vapor concentration at the interface. The difficulty of using Equation (2.2.2) is how one can express RH_s as a function of some known quantities.

The concept of the REA has incorporated the more conventional approach to expressing mass transfer at the boundary, i.e., Equation (2.2.1). It is an application of the chemical reaction engineering principle to establish a workable function for RH_s. In this approach, evaporation is modeled as zero order kinetics with activation energy, while condensation is described as the first order wetting reaction with respect to drying air vapor concentration without activation energy (Chen and Chen, 1997; Chen and Xie, 1997). The REA approach employs the Arrhenius equation in the evaporation term, which has an "origin" in the paper by Gray (1990), but is very different overall from what was proposed by Gray (Chen, 2008). The REA approach offers the advantage of being expressed in terms of simple, ordinary differential equations with respect to time. This negates the complications arising from use of partial differential equations (Chen, 2008). The REA does need accurate experimental data to determine model parameters, and an accurate equilibrium isotherm and surface area measurement. It can be shown later that the REA accommodates a natural transition because of the smooth activation energy as a function of moisture content (Chen, 2008). When activation energy for evaporation becomes higher than the latent heat of water evaporation, free water should have already been removed (Chen, 2008).

As a lumped approach (REA was initially proposed as a lumped parameter model), the drying rate (flux multiplied by surface area) of materials can be expressed as:

$$m_s \frac{d\bar{X}}{dt} = -N_c A = -h_m A\left(\rho_{v,s} - \rho_{v,\infty}\right) \quad (2.2.3)$$

where m_s is the dried mass of thin layer material (kg), X is the moisture content on dry basis ($kg \cdot kg^{-1}$) and $\bar{X}$ is the mean water content on dry basis ($kg \cdot kg^{-1}$), t is time (s), $\rho_{v,s}$ is the vapor concentration at the interface ($kg \cdot m^{-3}$), $\rho_{v,b}$ is the vapor concentration in the drying medium ($kg \cdot m^{-3}$), h_m is the mass transfer coefficient ($m \cdot s^{-1}$) and A is the surface area of the material (m^2). The mass transfer coefficient (h_m) is determined based on established Sherwood number correlations or established experimentally for the specific drying conditions involved (Incroper and Dewitt, 1990). Equation (2.2.3) is basically correct for all cases of water vapor transfer from a porous solid. In other words, there is no assumption of uniform water content in this approach, even though it was started with the mean water content. The surface vapor concentration ($\rho_{v,s}$) can be correlated in terms of saturated vapor concentration of water ($\rho_{v,sat}$) by the following equation (Chen and Chen, 1997; Chen and Xie, 1997):

$$\rho_{v,s} = \exp\left(-\frac{\Delta E_v}{RT_s}\right)\rho_{v,\text{sat}}\left(T_s\right) \quad (2.2.4)$$

where ΔE_v represents the additional difficulty of removing moisture from materials on top of the free water effect. It was thought that it would be excellent (and indeed lucky) to be able to relate ΔE_v to the average water content of the material. In other words, this ΔE_v is ideally to be moisture content (X) dependent. T is the temperature of the material being dried (K). For a small temperature range, say from 0°C to just over 100°C, $\rho_{v,sat}$ ($kg \cdot m^{-3}$) can be estimated using the following equation (Chen, 1998):

$$\rho_{v,\text{sat}}\left(T\right) = K_v \exp\left(-\frac{E_v}{RT}\right) \quad (2.2.5)$$

where K_v was found to be 1.61943×10^5 ($kg \cdot s^{-1}$) and E_v was found to be 38.99 $kJ \cdot mol^{-1}$. E_v is similar to the latent heat of water vaporization illustrating the physics involved (Chen, 1998). This is in line with the idea that evaporation is an activation process while condensation is not. The activation energy of the pure water evaporation reaction is equivalent to the value of the latent heat of water evaporation, as suggested earlier, based on classic physical chemistry.

For a wider temperature range, one can use the following, which correlates to the entire range of the data (0°C to about 200°C) summarized by Keey (1992):

$$\begin{aligned}\rho_{v,\text{sat}} &= 4.844\times10^{-9}(T-273)^4 - 1.4807\times10^{-7}(T-273)^3 + 2.6572\times10^{-5}(T-273)^2 \\ &\quad - 4.8613\times10^{-5}(T-273) + 8.342\times10^{-3}\end{aligned} \quad (2.2.6)$$

where T is temperature (K) based on the given data (Putranto *et al.*, 2010) (Figure 2.1.8).

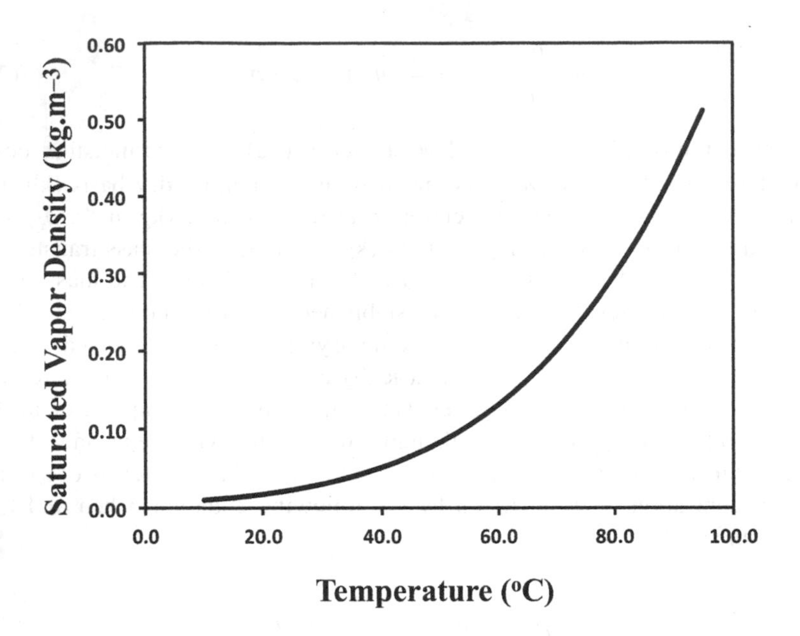

FIGURE 2.1.8 Saturated water vapor concentration in air under 1 atm (Equation 1.21).

The mass balance is then expressed as:

$$m_s \frac{d\bar{X}}{dt} = -h_m A\left(\exp\left(\frac{-\Delta E_v}{RT} \right) \rho_{v,\text{sat}}\left(T_s \right) - \rho_{v,\infty} \right) \tag{2.2.7}$$

For small objects such as particles or thin layer materials, the material temperature T is approximately the same as the surface temperature T_s. Basically, this happens when the Chen–Biot number is sufficiently small (Chen and Peng, 2005; Chen, 2007; Chen and Mujumdar, 2008). In this case, uniform temperature can be assumed throughout the material being dried such that one only needs to couple Equation (2.2.7) with a lumped energy balance to "govern" the drying process.

The activation energy (ΔE_v) is determined experimentally by rearranging Equation (2.2.7) as follows:

$$\Delta E_v = -RT_s \ln\left(\frac{-m_s \dfrac{dX}{dt} \dfrac{1}{h_m A} + \rho_{v,\infty}}{\rho_{v,\text{sat}}} \right) \tag{2.2.8}$$

where $d\bar{X}/dt$ is determined from experimental data on weight loss. It has been found, based on practical experiences of using the REA, that drying experiments to generate the REA parameters need to be conducted where the air (or gas) humidity is very low in order to cover the widest range of ΔE_v versus $\bar{X}$. The dependence of activation energy on moisture content can be normalized as:

$$\frac{\Delta E_v}{\Delta E_{v,\infty}} = \varsigma(X - X_\infty) \tag{2.2.9}$$

where ζ is a function of moisture content difference and $\Delta E_v \infty$ is at its maximum when the moisture concentration of the sample approaches relative humidity and temperature of the drying air:

$$\Delta E_{v,\infty} = -RT_\infty \ln(RH_\infty) \tag{2.2.10}$$

X_∞ is the equilibrium moisture content corresponding to RH_∞ and T_∞, which can be related to one another by the equilibrium isotherm (Keey, 1992). It is worth noting again that, so far, the experiments to achieve the relevant equation (Equation 2.2.9) have been conducted under very dry air conditions so the final water content attained is usually fairly low.

For the same material, the same initial water content and the same initial sample size (sometimes different sizes do not matter), the relationship (Equation (2.2.9)) may be viewed as unique, as many experimental results obtained under the above conditions but different drying conditions for the same material produced more or less the same trend quantitatively (Chen, 2008). This aspect will be shown later in various applications described in the forthcoming chapters.

The REA parameters for the drying of a material can be obtained from one good drying experiment and can then be applied to other different drying conditions (different drying air temperatures and air velocities) if the normalized activation energy would collapse to the same profile in these cases. However, the REA parameters should be generated from the material with same initial moisture content, since the activation energy has been found to be dependent on initial moisture content as well (Chen, 2008).

In many scenarios tested involving different materials, Equation (2.2.9) holds – a very pleasant outcome indeed. Of course, other forms of the equation are possible, and one should be worried if there was a temperature dependence function, or a material structure parameter became involved.

When the temperature of the moist material being dried does not vary much within itself, a uniform temperature may be considered (more quantitative assessment of this assumption can be found in a later part of this book that describes the modified Biot number and modified Lewis number). This leads to:

$$T_s \approx \overline{T}$$

where $\overline{T}$ represents the mean temperature of the material (K). For droplets and particles, this is satisfied basically.

This has allowed us to present the REA energy balance in a "lumped capacitance" (Incropera and Dewitt, 1990) for the material being dried:

$$mC_p \frac{d\overline{T}}{dt} = hA\left(T_\infty - \overline{T}\right) + H_{\text{drying}} m_s \frac{d\overline{X}}{dt} \tag{2.2.11}$$

where the mass of the material being dried is expressed as:

$$m = m_s\left(1+\overline{X}\right) \tag{2.2.12}$$

2.2.2 Application of the REA to Describe Droplet Drying Kinetics

The REA has been used to model drying of a range of food materials such as pulped kiwifruit leather, whey protein concentrate, lactose, skim milk powder, whole milk powder, cream, and mixtures of sugars (Chen *et al.*, 2001; Chen and Lin, 2005; Lin and Chen, 2005, 2007). Results showed that this approach models moisture content and temperature profile versus drying time very accurately. For example, modeling of drying of aqueous lactose solution droplets showed that the average absolute difference of weight loss profile was about 1% of the initial weight, while that of the temperature profile was about 1.2°C. Moreover, application of the REA to model drying of cream and whey protein concentrate showed average absolute errors of weight profiles of 1.9% and 2.1% respectively, while errors for temperature were about 3°C and 1.9°C respectively (Lin and Chen, 2007). Modeling of skim milk and whole milk powder by the REA was also accurate (Chen and Lin, 2005).

The REA has been implemented in computational fluid dynamics (CFD)-based simulations of spray dryers to couple the dispersed phase (droplets dried) and the continuous phase (drying air) (Woo *et al.*, 2008b; Jin and Chen, 2009a,b, 2000). CFD simulations using the REA can predict outlet air temperature and outlet particle moisture content reasonably well compared with the experimental data. In addition, the REA was also implemented to predict the evaporation zone, drying rate, trajectory of particles, and deposition of particles in spray dryers (Woo *et al.*, 2008b). The application of CFD in conjunction with the REA to describe the performance of industrial-scale spray dryers in 2-D and 3-D was conducted (Jin and Chen, 2009a,b). Patel *et al.* (2009a) have extended the "single solid component" approach of the REA to a composite REA model – drying kinetics for mixtures of "non-interacting" solutes (maltodextrin and sucrose). The activation energy of the mixture was determined based on the mass fraction of each solute and their corresponding activation energies. It was shown that the average relative error between experimental and calculated data was below 1.5% for droplet weight and below 3% for droplet temperature (Patel *et al.*, 2009a).

2.3 CHARACTERISTIC DRYING CURVE MODELS

2.3.1 Theoretical Framework

The characteristic drying curve (CDC) model is another approach that can be used to arrive at a unified behavior on how the falling rate of the droplet changes throughout the drying process (Langrish and Kockel, 2001). In the model, the highest possible drying rate, the drying rate corresponding to the wet bulb period of evaporation, is taken as the reference point from which the drying rate is progressively reduced. This highest possible drying rate is given as follows:

$$-\frac{dm_w}{dt} = \frac{h A_p \left(T_b - T_{p,wb}\right)}{\Delta H_L} \tag{2.3.1}$$

Taking this as the upper limit of drying, one would then expect the drying rate to progressively reduce in the secondary drying period. The moisture content of the droplet can then be arbitrarily adopted to delineate the extent to which the drying has progressed. Secondary drying only begins once the droplet reaches critical moisture content, and the lowest possible moisture content is the equilibrium moisture content. Therefore, these two moisture contents can be adopted to normalize the progressively reducing moisture content of the droplet to arrive at a dimensionless reduction factor as follows:

$$f = \left[\frac{X - X_{eq}}{X_{cr} - X_{eq}}\right]^N \tag{2.3.2}$$

It is noteworthy that at the initial instance of the falling rate period, the reduction factor will be unity. As the drying rate drops when the droplet progresses into the falling rate period, this ratio then drops and eventually reaches zero when the droplet approaches equilibrium moisture content. Combining both equations, the CDC approach to describe the falling rate drying behavior then becomes:

$$-\frac{dm_w}{dt} = \left[\frac{X - X_{eq}}{X_{cr} - X_{eq}}\right]^N \frac{h A_p \left(T_b - T_{p,wb}\right)}{\Delta H_L} \tag{2.3.3}$$

This arbitrary form for the reduction factor is mathematically convenient, as the value N then denotes the shape or, if physically interpreted, the behavior in which the solidification affects the progressive drying retardation. A value of 1 denotes a linear falling rate behavior. A value of greater than 1 for N implies a sudden retardation in the drying rate, most likely due to a significant skin-forming phenomenon. While this mathematical profile for the reduction factor hitherto fits commonly observed drying behavior so far, it is mainly arbitrary; it can be expressed in other forms if it is suitably normalized.

2.3.2 How to Obtain the Parameter N from Experiments

Different methods can be used to firstly obtain the data on how the mass of the sample of interest changes over the drying time. This can be from thin film or single droplet or another form of experimental technique. By computing the gradient at each mass data throughout the drying process, a plot of drying rate versus droplet moisture content can then be obtained (Figure 2.3.1). Normalize the drying rate as follows, which is a rearrangement of Equation (2.3.3):

$$\frac{-\dfrac{dm_w}{dt}}{hA\left(T_b - T_{p,wb}\right) / \Delta H_L} = \left[\frac{X - X_{eq}}{X_{cr} - X_{eq}}\right]^N \tag{2.3.4}$$

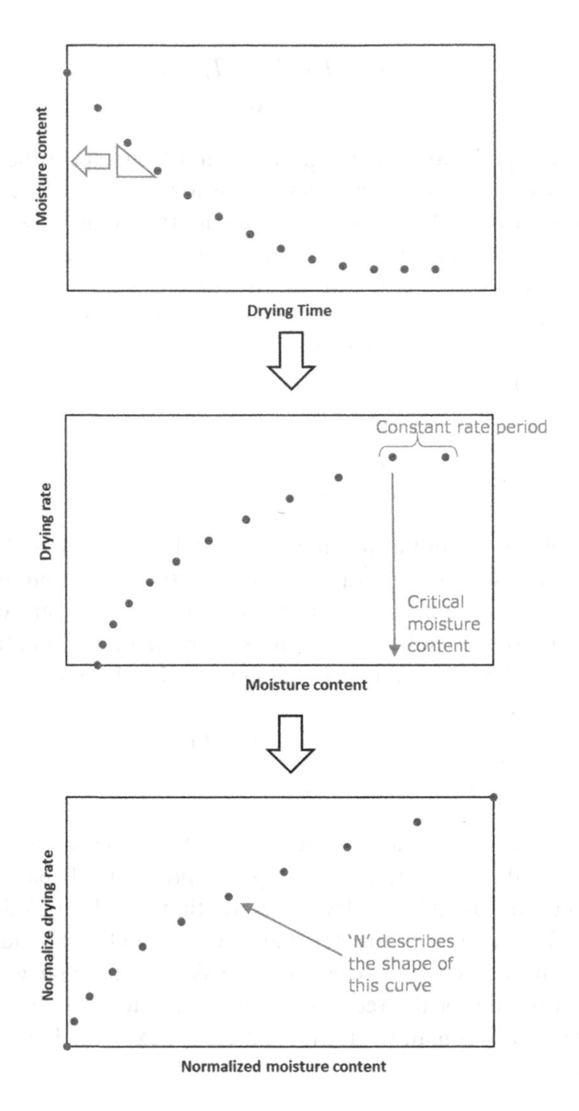

FIGURE 2.3.1 Obtaining the CDC drying kinetics.

The highest possible experimentally measured drying rate should also correspond to the wet bulb drying rate for low initial droplet solid concentration. The critical moisture content can be obtained by examining the moisture content when the drying rate starts to reduce. The equilibrium moisture content can be obtained from available isotherm data.

This procedure should then be repeated by drying the material under different temperatures, humidities, and velocities (different drying conditions) if the experimental rig permits. The normalized data described above from each experimental run can then be collated and the parameter N can then be obtained by fitting using only the set of normalized data at normalized moisture content lower than unity, delineating when the particles have entered the secondary drying period.

TABLE 2.3.1
Characteristic Drying Curve Parameters for Some Common Spray Dried Materials

Materials	References	CDC index	Critical moisture content
Skim milk	Langrish and Kockel (2001)	1 (linear)	Initial moisture content (80% wt analyzed)
Sucrose	Woo *et al.* (2008a)	2.58	Initial moisture content (50% wt analyzed)
Maltodextrin	Woo *et al.* (2008a)	3.22	Initial moisture content (50% wt analyzed)
Sucrose–maltodextrin (1:1)	Woo *et al.* (2008a)	1.98	Initial moisture content (60% wt analyzed)
Maltodextrin	Zbicinski and Li (2006)	1 (linear)	From pilot scale conditions: Evaporation rate >8 g/cm^{2}s (10% wt initial moisture) Evaporation rate >5 g/cm^{2}s (30% wt initial moisture) Evaporation rate >3 g/cm^{2}s (50% wt initial moisture) The critical moisture content was close to the initial moisture

Source: Woo, M.W. *Computational Fluid Dynamic Simulation of Spray Dryers – An Engineer's Guide*. CRC Press, Boca Raton, FL, 2016. With permission.

2.3.3 Compilation of Falling Rate Curves

The CDC index for some commonly spray dried materials is given in Table 2.3.1. For most applications of the CDC model, the initial moisture content is typically taken as the critical moisture content. The simplifying assumption here is that the initial heating-up time and the constant drying period are very short and rapid relative to the overall spray drying process. It is important to note that the critical moisture content, fundamentally, may be dependent on the initial concentration of the feed solution and the drying rate of the droplets (Zbicinski and Li, 2006).

2.4 PREDICTION OF SURFACE COMPONENT COMPOSITION, MIGRATION, AND PACKING

2.4.1 The Need for a Multiscale Model for Surface Composition Prediction

The spray drying process, together with the feed material, determines the micro- to macroscale structures of spray dried particles, and hence their properties and functionalities (see Figure 2.4.1). The surface structure and composition are especially important, because they significantly influence particles' properties (such as

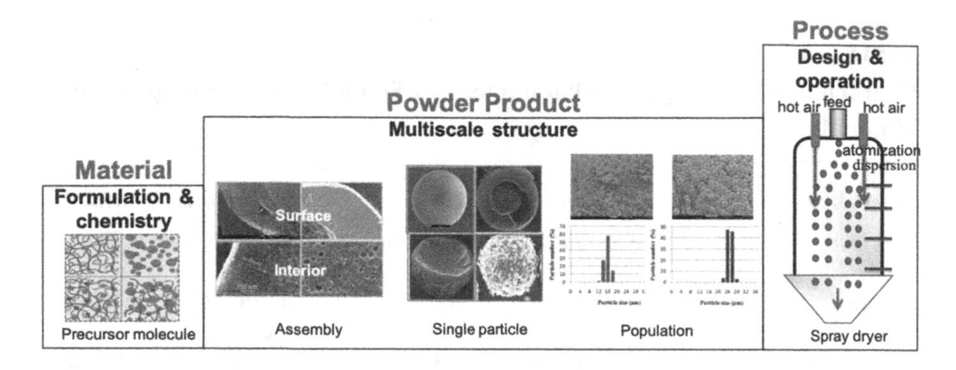

FIGURE 2.4.1 Multiscale view of the spray drying system.

wettability and stickiness) as well as their dissolution or rewetting behavior (Vehring *et al.*, 2007; Wu *et al.*, 2014). Consequently, reliable prediction of surface composition becomes crucial for powder quality control.

By resorting to electron spectroscopy for chemical analysis (ESCA, also known as X-ray photoelectron spectroscopy (XPS)), many experimental investigations have reported a big difference between the surface composition and the bulk composition of spray dried particles. This observation suggests that solute segregation must occur during spray drying (Kim *et al.*, 2003; Fäldt and Bergenståhl, 1994, 1996a, 1996b).

The distributed parameter model developed by Langrish and his group (Wang and Langrish, 2009; Wang *et al.*, 2013) is one representative effort to characterize solute segregation. In order to obtain localized composition, a particle was divided into a number of shells. The diffusive mass flux of each solute component was calculated individually for each layer before being combined to estimate the total mass flux. However, the predicted surface concentration was always lower than that calculated from the XPS measurements. Another representative approach was proposed by Chen *et al.* (2011). In their work, the governing equations of convective diffusion were simplified into a set of algebraic equations by using the idea of the characteristic length. The solute composition on the particle surface was then obtained by solving those algebraic equations. For a two-component system (i.e., the protein–lactose system), the theoretical estimates, however, were always greater than the XPS results.

The methods listed above are both based on the continuum theory. It is understandable that their predictions cannot match XPS results on surface composition. As shown in Figure 2.4.2, the surface layer in a continuum model is the layer α (not drawn to scale), whose thickness corresponds to the mesh grid size used in solving the governing equations of mass transfer. The XPS detectable surface layer, γ, has a thickness of ~10 nm only, which is even smaller than the size of many solute molecules (such as protein and fat). The continuum theory is not capable of characterizing solute segregation and powder surface formation at nano-scale, and hence this theory alone cannot predict surface composition that needs to be validated by the XPS data.

2.4.2 Key Ideas and Calculation Procedure

The continuum mass transfer model has to be extended so that it can take care of molecular packing in the surface layer α. A variable characterizing the packing status has to be introduced, i.e., the packing fraction ϕ_s. This variable gives the volume fraction of all solute molecules in the α layer. One can use diffusion theory to quantify both the packing fraction (i.e., the volume fraction) and the surface coverage (i.e., the mass fraction ω). At the same time, by resorting to the geometric packing theory, one can derive the relationship between the packing fraction and the surface coverage for spherically shaped solute molecules. The equation set obtained from the two sets of theories can be solved together to obtain ϕ_s and ω. By assuming different packing schemes, one can further quantify the surface coverage in the γ layer (see Figure 2.4.2).

For the surface composition of spray dried two-component particles, the calculation procedure is summarized below:

Step 1. A number of input data have to be given, which include material properties and system parameters, such as the radius of a molecule of solute i (R_i), the solute density ρ_i, and the XPS detectable depth h, etc.

Step 2. Given the overall packing fraction ϕ_s, solve the diffusion model (Equations (2.4.1) and (2.4.2)) to obtain the surface composition $c_{1,s}$ and $c_{2,s}$.

$$\frac{R_1}{R_2} = \frac{C_{1,s} - C_{1,o}}{C_{2,s} - C_{2,o}} \cdot \frac{2C_{2,o} + (\beta + \xi)(C_{2,s} - C_{2,o})}{2C_{1,o} + (\beta + \xi)(C_{1,s} - C_{1,o})} \tag{2.4.1}$$

$$\phi_s = \frac{C_{1,s}}{\rho_1} + \frac{C_{2,s}}{\rho_2} \tag{2.4.2}$$

where $c_{1,o}$ and $c_{2,o}$ ($\text{kg} \cdot \text{m}^{-3}$) are respectively the initial concentrations of solute 1 and solute 2 in the droplet before drying. The mass transfer affected region can be assumed to be 25% of the droplet radius, i.e., $\beta \approx 0.25$.

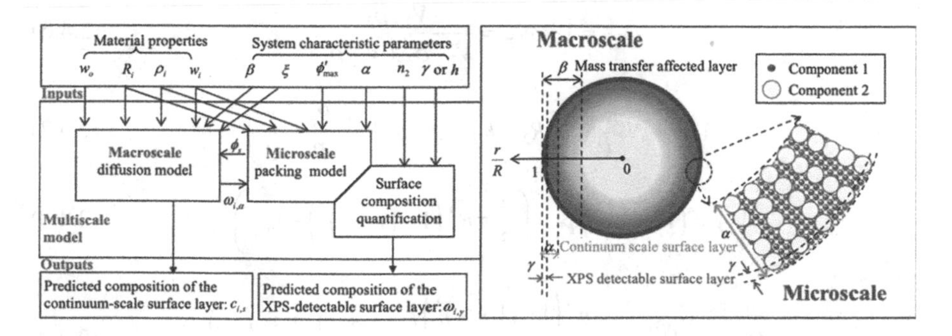

FIGURE 2.4.2 Model framework for surface composition prediction. The key is to couple the macroscale diffusion model with the microscale geometric packing model. (Adapted from Xiao, J., Chen, X.D, Multiscale modeling for surface composition of spray-dried two-component powders. *AIChE J.* 60, 2416–2427, 2014.)

The radial velocity of solvent is close to the shrinking velocity of the droplet surface, i.e., $\xi \approx 1$. Plug $c_{1,s}$ and $c_{2,s}$ into Equation (2.4.3) to obtain the surface coverage $\omega_{2,\alpha}$.

$$\omega_{2,\alpha} = 1 - \omega_{1,\alpha} = \frac{c_{2,s}}{c_{1,s} + c_{2,s}} \tag{2.4.3}$$

Repeat this step for different values of ϕ_s, and then plot $\omega_{2,\alpha}$ as a function of ϕ_s. Note that $\omega_{2,\alpha}$ gives surface coverage in the α layer, which is much thicker than the XPS detectable layer.

Step 3. For a specific surface coverage $\omega_{2,\alpha}$, solve the microscopic packing model (Equations (2.4.4) and (2.4.5)) to obtain the overall packing fraction ϕ_s.

$$\varphi_s = \varphi_1 + (1-\varphi_1)\left(1 + \frac{\rho_2}{\rho_1}\frac{1}{\varphi_1}\frac{1-\omega_{2,\alpha}}{\omega_{2,\alpha}}\right)^{-1} \tag{2.4.4}$$

$$\begin{cases} \varphi_1 = \min\left(\varphi'_{\max},\ \dfrac{\omega_{1,\alpha}}{\omega_{2,\alpha}}\dfrac{\rho_2}{\rho_1}\left(\dfrac{1}{\varphi'_{\max}}\max(1,\ \psi) - 1\right)^{-1}\right) \\ \psi = \dfrac{1-(1-\alpha)^3}{2-\varphi'_{\max}}\left(1 + \dfrac{w_1}{w_2}\dfrac{\rho_2}{\rho_1}\right) \end{cases} \tag{2.4.5}$$

The maximum packing volume fraction of perfect uniformly sized spheres is $\phi'_{\max}$ (i.e., 0.631) based on the Quemada model.

Repeat this step for different values of $\omega_{2,\alpha}$, and then plot ϕ_s as a function of $\omega_{2,\alpha}$.

Step 4. Two curves obtained from both models can be plotted in one figure to identify their intersection point, which offers the solutions of ϕ_s and $\omega_{2,\alpha}$.

Step 5. Calculate the surface coverage in the XPS detectable layer using the following equations:

$$\omega_{2,\gamma} = 1 - \omega_{1,\gamma} = \frac{V_{2,\gamma}\rho_2}{V_{2,\gamma}\rho_2 + V_{1,\gamma}\rho_1} \tag{2.4.6}$$

$$V_{2,\gamma} \approx N\frac{\pi}{3}h^2(3R_2 - h) = N\pi\gamma^2 R^3\left(\frac{R_2}{R} - \frac{\gamma}{3}\right) \tag{2.4.7}$$

$$V_{1,\gamma} = (V_\gamma - V_{2,\gamma})\varphi_1 = \left(\frac{4}{3}\pi R^3\left(1-(1-\gamma)^3\right) - V_{2,\gamma}\right)\varphi_1 \tag{2.4.8}$$

$$N = \frac{V_\alpha \varphi_2}{n_2 \frac{4}{3}\pi R_2^3} = \frac{R^3\left(1-(1-\alpha)^3\right)}{n_2 R_2^3}\left(1 + \frac{\rho_2}{\rho_1}\frac{1}{\varphi_1}\frac{1-\omega_{2,\alpha}}{\omega_{2,\alpha}}\right)^{-1} \tag{2.4.9}$$

where the ratio of the α layer thickness to the droplet radius is α and the droplet radius is R.

2.4.3 The Capabilities and Limitations of the Method

The approach described above was introduced in detail in Xiao and Chen (2014). The methodology has been successfully utilized to study lactose–protein two-component systems. The effect of protein addition into the lactose solution on the final powder surface composition demonstrated by XPS data can be appropriately captured, even for the very low end of the protein bulk concentrations. Although the exact packing scheme cannot be obtained by this approach, the proposed extreme schemes can offer lower and upper bounds of the surface coverage. An interesting phenomenon shown in Figure 2.4.3 is that adding a very small amount of protein into the lactose solution can lead to drastic increase of surface coverage of protein for the spray dried powders. This feature can be successfully predicted by the multiscale approach, which clearly outperforms the mono-scale diffusion model, since the latter cannot even capture the correct trend of the XPS data. The results also confirm that the surface coverage values are different for the α layer and the γ layer. However, they have conventionally been treated by many researchers as the same physical quantity, leading to prediction errors and misinterpretation of XPS results.

The multiscale model was further extended to take into account the size distribution of atomized droplets by Xiao *et al.* (2016). The method was no longer restricted

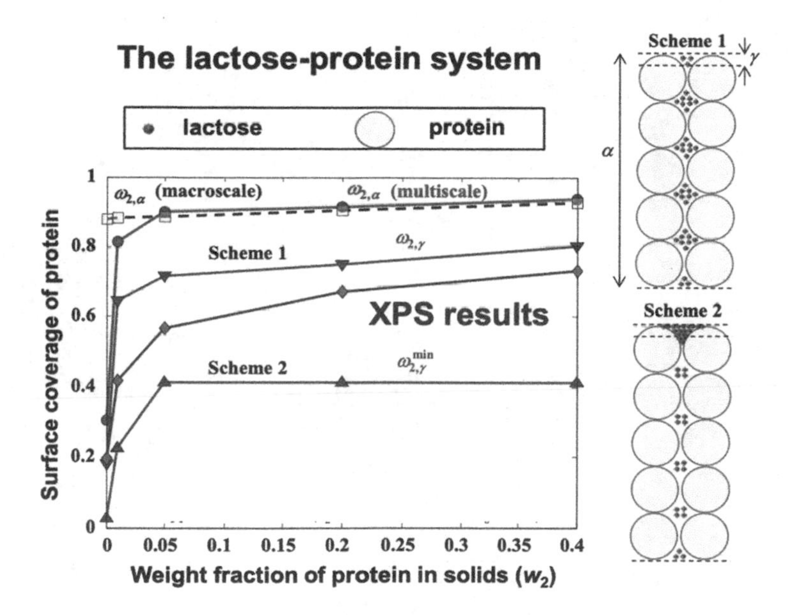

FIGURE 2.4.3 The surface coverage of protein as a function of the initial solution compositions for a lactose–protein two-component system: comparison between the experimental XPS results by Fäldt and Bergenståhl (1994), theoretically predicted results using the continuum diffusion approach by Chen *et al.* (2011), and the results predicted by the multiscale approach (species 1 is lactose, species 2 is protein, $n_2 = 5$, $\alpha = 1\%$, $h = 10$ nm, constant w_o [i.e., 10% solid content] and variant w_2). Lactose molecules distribute uniformly in scheme 1, while they aggregate towards the surface in scheme 2. (Adapted from Xiao, J., Chen, X.D., Multiscale modeling for surface composition of spray-dried two-component powders. *AIChE J.* 60, 2416–2427, 2014.)

to the study of uniform-sized droplet drying. The power of the method was also demonstrated through the new molecular level interpretation of XPS data. This can be used to rationally explain the difference between the surface composition predicted by the distributed parameter model and the XPS measurements (Xiao *et al.*, 2015).

The current multiscale approach, however, cannot offer surface structure of the particles. The drying dynamics cannot be characterized, and the method is currently restricted to the study of two-component systems. In future, microscale models such as Monte Carlo or molecular dynamics models that can describe the evaporation-induced self-assembly of more than two types of solutes should be developed. In this way, three-dimensional structure evolution for powder surface formation during the drying process can be quantitatively described. It will play a critical role in structure-focused powder quality control.

3 Spray Dryer Designs Including Multistage

3.1 CO-CURRENT VERSUS COUNTER-CURRENT

Most of the commercially available spray dryers for food application, in essence, are co-current spray dryers. Despite the air eventually recirculating upwards once it reaches the bottom static bed (if available), the spray and the hot drying air enter the chamber in the same "direction" within the drying chamber. An alternative configuration would be to employ counter-current spray drying, in which the introduction of the hot air and the spray occur in opposite directions. There are not many reported food applications using this form of dryer. At the moment, this form of spray drying is mainly used for the production of non-heat-sensitive materials such as detergents. More information on the equipment configuration of both forms of dryer can be found in other handbooks available in the literature (Filkova *et al.*, 2015; Masters, 1979). For counter-current spray tower configurations, in which the spray is introduced at the top while the hot air enters from the bottom, more specific details can be found in Wawrzyniak *et al.* (2017) and Ali *et al.* (2017). An inherent limitation in this form of counter-current tower, which will be discussed in detail later on, is the potential entrainment of the sprayed droplets or powder. Alternatively, counter-current drying can also be introduced by upward spraying of the feed material against a downward flow of hot air; the powder will then eventually flow downwards following the airflow.

The primary difference between co-current and counter-current spray drying is the drying history experienced by the droplet. These differences are summarized in Table 3.1.1. In co-current spray drying, the droplets will initially be in contact with the relatively lower-humidity and higher-temperature hot air. This will then induce very rapid solidification, typically quenching the droplet into amorphous forms of powder. Despite the contact with high-temperature hot air, the rapid evaporation will also maintain the droplet at the wet bulb temperature, minimizing thermal damage to the droplet. This removal of moisture will also rapidly cool the hot air. Therefore, the droplet will experience progressively lower temperature throughout the chamber. It is widely accepted that these two factors will help to preserve the quality of heat-sensitive products despite the high inlet hot air temperature used in spray drying. The final state of the powder will then be determined by the humidity and temperature of the air leaving the chamber.

In counter-current spray drying, however, the droplet will initially contact relatively cooler drying air as, due to the counter-current arrangement, the incoming hot air will have been cooled by the evaporating droplets. Inevitably, the droplets will also initially be in contact with relatively more humid air. They will then experience progressively hotter air. This means that even if the droplets experience the wet bulb

TABLE 3.1.1
Summary of the Differences between Co-Current and Counter-Current Spray Drying ("High" and "Low" are Mainly Indicative Comparison)

Drying characteristics	Co-current spray drying	Counter-current spray drying (top-down spray)	Counter-current spray drying (upward fountain spray)
Temperature of initial air contact	High	Low	Low
Humidity of initial air contact	Low	High	High
Rate of initial solidification	High	Low	Low
Temperature of air in latter stages of drying	Low	High	Low
Humidity of air in latter stages of drying	High	Low	High

temperature, there will be a progressive increase in the wet bulb temperature corresponding to the progressively hot air temperature. In contrast, as the initial contact is at relatively lower temperature, one would expect less rapid solidification of the droplets. There are reports in the literature to indicate that this may be beneficial in the control of in-situ crystallization, which required more time for functional structure formation during the trajectory of the droplets in the chamber. Depending on the configuration of the counter-current tower, if the spray was introduced from the top (downwards), the condition of the powder is then not dependent on the outlet condition of the air leaving the chamber, in contrast with co-current spray drying. In other words, the powder will be leaving the chamber while in contact with hotter and dryer air, which will allow the potential for a dryer powder to be produced. On the flip side, if fountain-like spray drying is used, the droplets will experience the drying history of a counter-current spray dryer while having the final powder conditions determined by the outlet air conditions, similarly to the co-current dryer.

One big potential advantage of counter-current spray drying is a more energy-efficient operation of the process. This is simply due to the nature of the counter-current procedure, which maximizes the temperature difference (and the progressively lower humidity) between the dehydrating air and the droplet for heat and mass transfer. For the configuration in which the spray is introduced from the top with the air entering the chamber from the bottom, we have undertaken a theoretical exploration to show that the energy usage of the spray dryer can be reduced significantly. Due to the efficient heat and mass transfer, we have also shown that there is potential to reduce the drying temperature to below 100°C (Razmi *et al.*, 2019). To achieve this, there needs to be a delicate balance between the initial sprayed droplet size and the mass flow rate of the air into the drying chamber. This is important to minimize powder entrainment. Another point is to ensure that the progressively cooled outlet air temperature is not lower than or does not approach the wet bulb temperature, the theoretical limit of counter-current spray drying. It is noteworthy that the theoretical limit of co-current spray dryers (possible evaporation rate relative to the mass flow rate of the air) will be lower than that of counter-current dryers. More work is currently underway to fully and experimentally evaluate this potential configuration.

Despite this potential advantage, the counter-current spray dryer is not widely used for food powder applications. The concern in the industry is that the progressively higher air temperature may impart thermal damage to the food powder. This may be exacerbated by the progressively diminishing wet bulb effect as the droplet solidifies. More work will be needed to fully evaluate whether more efficient drying will allow the particle to rapidly traverse past the "critical moisture" period for thermal degradation. Critically reviewing the current practice in counter-current spray drying, there may be other factors in addition to the progressively higher air temperature that have contributed to this concern so far. Firstly, most counter-current spray drying reported so far has concerned detergent-based spray drying, which is not thermally sensitive, and therefore, there was no need to maximize evaporation efficiency with the aim of lowering the inlet air temperature. Therefore, the potential to convert counter-current spray drying into a low-temperature spray drying process has not been explored in the literature. Secondly, most of the counter-current spray drying trials reported in the literature involved the use of high swirling flows (Wawrzyniak *et al.*, 2017; Ali *et al.*, 2017). Although this improves the heat and mass transfer efficiency, this strategy will induce significant wall deposit layers (Francia *et al.*, 2015). It was found that these deposit layers, when collected together with the spray dried product, contribute to the "browning" of the product. Therefore, more work will need to be undertaken to determine if the browning from counter-current spray dryers, in general, stems from the deposit or the collected products themselves.

3.2 MULTISTAGE DRYING PROCESS AND FINES RETURN

Figure 3.2.1 shows a typical three-stage spray drying system. The first stage, which is the primary drying chamber, involves converting the atomized droplets into dried particulates or powder. Most operations will aim to achieve a powder moisture content of less than 10% wt (typically around 5% wt). This is mainly as a balance to achieve non-sticky powder, so that the powder will not plug up the internal static bed, yet will not require excessive high-temperature air to further reduce the moisture content within the short drying time in the primary drying chamber. The internal static bed is mainly an extension of the bottom section of the spray drying chamber with the main purpose of providing a longer residence time to remove the residual moisture content to about 2% wt. The longer residence time in the internal static bed will then allow the use of lower-temperature drying air at this second stage of drying. The airflow rate and the temperature used in the internal static bed will also need to be delicately balanced to ensure sufficient secondary drying but prevent excessive breakage of the agglomerates from the primary drying stage.

The third stage of drying is the external vibrating fluidized bed. Operation of the external bed is mainly a combination of drying (at the start of the bed) and cooling (progressively towards the end of the bed). For such applications, the temperature of the drying air entering the bed may be individually controlled. For products requiring further enhanced instantization, lecithin may also be sprayed or introduced

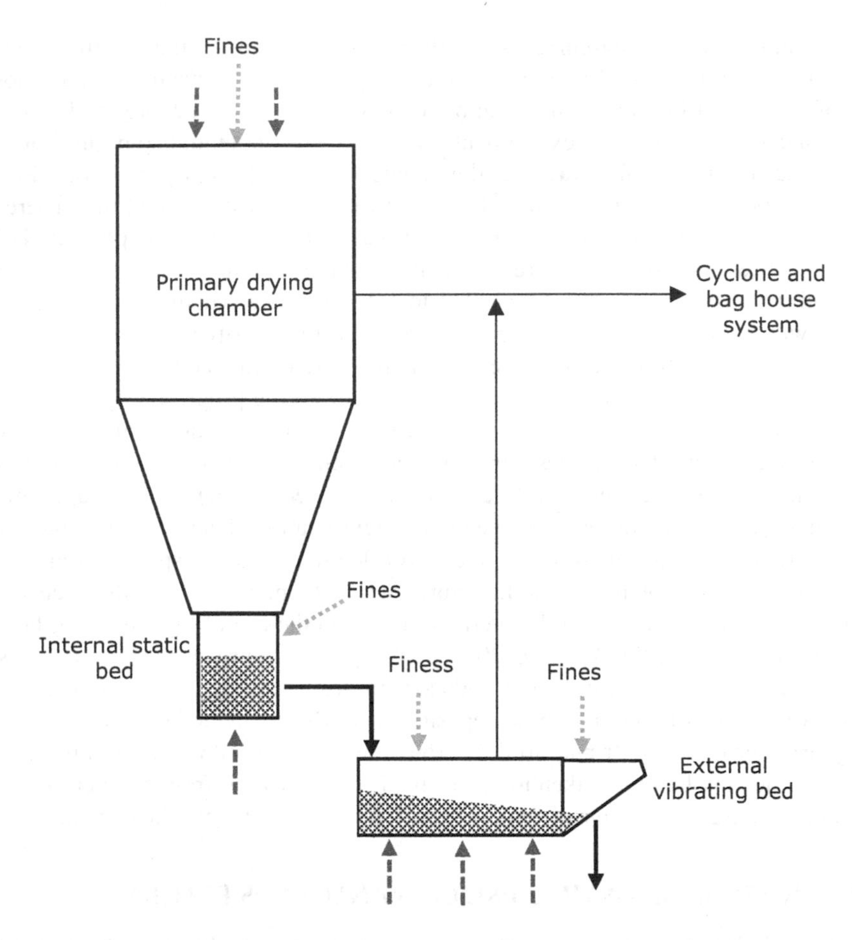

FIGURE 3.2.1 Typical three-stage spray drying process. (Adapted from Woo Advances in production of food powders by spray drying. In *Advanced Drying Technologies for Food*, edited by Mujumdar, A.S., Xiao, H.W. Taylor & Francis, Boca Raton, 2019.)

within this third stage onto the vibrating powder. In certain plant configurations, lecithin may even be introduced at the outlet of the vibrating fluidized bed.

With these three stages forming the process, there can be many strategies in which the fines generated from the process can be managed. Fine powder is carried away from the primary and third stage drying chamber via the exhaust air. The vacuum suction for the whole exhaust line is typically generated by a main vacuum blower after the bag house. The balance between the suction pressure and the air flow rate in the primary drying chamber is delicately balanced to ensure a slight negative pressure. This is mainly as a safety measure in the event of potential explosion. The suction from the external vibrating fluidized bed, however, can typically be controlled by a manual valve positioned in the outlet line from the external bed. This provides control of the amount of fine exhausts from the external vibrating bed.

Fines returned to the top section of the primary chamber are mainly intended to induce forced agglomeration with the atomized and partially solidified powder from the nozzles. While this will provide the most direct approach to control the final product size distribution and the agglomerated structure, it also exposes the fines to relatively high temperature in the primary drying chamber. An alternative is to then return the fines to the internal static bed. As the powder in the internal static bed is relatively dry and more solidified, the formation of agglomerates and the strength of the agglomerates formed may be less significant when compared with returning the fines to the top of the primary chamber. The return of fines to the external fluidized bed may not be expected to contribute significantly to the formation of a "forced" agglomerated structure; it is mainly to return the fines into the product for production yield management.

Regardless of the approach to the generation of fines and the return strategy, one main area requiring further development is in the measurement of the rate of fines being transported. At the time of preparing this book, the authors are not aware of any commercially adopted sensor or measuring device suitable for such an application in the industry. The management and control of fines, which to a certain extent may also affect agglomeration control, relies significantly on the operators' "feel" and experience.

3.3 ATOMIZATION

3.3.1 Selection and Balance between Viscosity and Concentration

Atomization refers to the process of breaking up a bulk liquid into droplets (Masters, 1979; Chen and Mujumdar, 2008). Common home atomizers are shower heads, garden hoses, hair sprays, or perfume sprays, etc. A spray is a "cloud" of moving droplets that move in a controlled manner. A droplet is a liquid particle usually of more or less a spherical shape. The reason for its roundness is due to the liquid surface tension. Bulk liquid has to be made into thin liquid ligaments that can then be torn into a massive number of small entities such as liquid particles. Water has a surface tension of 0.073 $N \cdot m^{-1}$ at 20°C (compared with that of mercury: 0.465 $N \cdot m^{-1}$ at the same temperature). Surface tension tends to stabilize a liquid, preventing it from being broken up into small sizes. As such, if all other factors are maintained the same, increasing surface tension increases the average droplet size. Liquid viscosity has a similar effect; the greater it is, the larger the average droplet size. Higher viscosity gives a resistance to the liquid entity from being torn apart as well. Liquid density is also a parameter to consider. It makes liquids resist acceleration (perhaps), so a greater density usually leads to larger average droplet sizes. The concentration and the temperature of a liquid such as milk dictates the corresponding liquid density. A review and compilation of suitable models to describe the viscosity of liquids due to changes in temperature and concentration can be found in Chapter 2 of Quek (2011).

This solute–solid concentration within the liquid feed, which increases both the density and the viscosity (though reduces the surface tension somewhat), is a dominant parameter in spray drying industries. Practically, one would also find that the greater the throughput (capacity) from each atomizer (nozzle or disk) of the same size, the greater the average droplet size. The conventional atomization processes for large-scale product usually include pressure (airless), air spray, and centrifugal (disk). However, in recent times, electrostatic and ultrasonic nozzles are also in fashion for some applications. Of course, the conventional ones can have capacities of several to tens of tonnes of product per hour. Even in the newer processes, the above arguments on the effects of surface tension, viscosity, density, and throughput are still valid.

3.3.2 What Is the Initial Droplet Size Distribution?

One common challenge or ambiguity faced by the industry is in determining the actual size distribution of the spray. Most of the commercially available atomizers are typically rated or characterized with water. In the dairy industry in particular, highly concentrated dairy sprays are typically used at concentrations of greater than 50% wt. The strategy is mainly to reduce the cost of evaporation, making production economically more efficient. Such high-concentration spray will differ significantly in terms of viscosity, density, and concentration when compared with water. It is well known that these properties will significantly affect the atomization behavior of nozzles. Therefore, suitable nozzle operating conditions will have to be determined by trial and error. There are, however, correlations available in handbooks, which can be used to link these feed and atomization parameters to the droplet size distribution. From the author's experience in using these correlations, it may be useful to use the correlations as a guide to inform how the different parameters affect the atomization behavior in a relative manner. Due to the myriad forms of atomizer (nozzles) available, usage of these correlations as an absolute determination of initial droplet size distribution may not be the most accurate approach.

While suitable initial droplet size distributions can eventually be determined from trial and error in actual operation, lack of this information also limits the accuracy of any modeling or predictive work for spray drying. From some analysis undertaken (unreported), even a variation of ±30 microns in the specification of the initial droplet size will significantly affect the overall prediction of droplet drying behavior and its subsequent trajectory within the spray drying chamber. In a way, this remains one of the main ambiguities limiting the accuracy possible in the modeling of industrial-scale dryers.

4 Mass and Energy Aspects of Spray Drying

4.1 ZERO DIMENSION MODELING OF SPRAY DRYERS

4.1.1 EQUILIBRIUM OUTLET MOISTURE-BASED MODEL

This type of predictive tool treats the spray drying chamber as a gray box and provides only the outlet conditions from it. The primary advantage of this form of spray dryer modeling, which in essence is a mass and energy balance for the operating unit, is to allow plant-wide prediction of the drying process in the steady state. One of the main challenges in such a model is in the prediction of the final powder moisture content leaving the drying chamber. Ozmen and Langrish (2003) suggested assuming that the powder is in equilibrium with the outlet air thermal and humidity conditions. This is a well-accepted assumption in view of the large size of the spray drying chamber in which the powder is expected to have sufficient residence time to reach or approach equilibrium given the air conditions. Although not explicitly mentioned, this may also contribute to the fact that the outlet air conditions are representative of the overall conditions within the drying chamber. With this assumption as the backdrop, based on the conditions at the inlet to the drying chamber, the outlet conditions can then be arrived at by successive iterations between the powder condition (moisture content) and the air thermal and humidity conditions. The sorption isotherm of the product is commonly used to provide a link between the air and powder conditions.

Some reported measurements in industrial-scale operations, however, reported deviations from this equilibrium assumption. Straatsma *et al.* (1991), based on industrial trials and measurements, reasoned that the powder would not have sufficient time to arrive at the equilibrium value described by the material-specific isotherms. Their approach is to fit, based on trial measurements, a parameter to calibrate the predictive framework to arbitrarily increase the powder moisture content prediction relative to the equilibrium values. For future workers adopting this method, it will be important to verify the position of the powder sampling, as the spray dried powder may spend a significant amount of time in the conveying or cyclone system in large-scale multistage systems.

Schuck *et al.* (2009) used a desorption technique, SD²P®, to measure the extra energy requirement for drying, making reference to the evaporation of pure water. Using the psychrometric chart as a basis, they proposed a method to incorporate this additional energy into the mass and energy balance of a spray drying chamber. Deviating from the two methods described earlier, the SD²P® method computes the inlet conditions while keeping the outlet conditions fixed. More details will be discussed in subsequent sections of this book.

Dynamic simulation of such gray box simulation of the spray drying chamber can be further developed for control of simulation of the process, particularly to investigate how the spray drying system handles disturbances in operation or fluctuations in the ambient conditions. Petersen *et al.* (2017) proposed such a dynamic mass and energy model incorporating the internal static and external bed in the overall drying process. The proposed model, however, relies on a large number of fitted transfer coefficients.

4.1.2 SD²P® and the Slow Desorption Approach by INRA

A desorption method based on a thermodynamic approach has been developed to evaluate the behavior of dairy concentrates during drying. Involving overall heat and mass balance throughout a spray dryer, this approach can determine several key gas-feed parameters for industrial spray drying processes. A specifically designed spray drying software package (SD²P®) was then designed following this approach to predict the optimal inlet drying air temperatures with acceptable accuracy (95%–99% accuracy) for spray drying of dairy products. This is also called the "drying by desorption method", and the SD2P® software was established successfully at the Institut National de la Recherche Agronomique (INRA) in France (Schuck *et al.*, 2009).

Essentially, comparing against the evaporation of pure water under the same drying conditions, if one knows how much energy would be needed for drying concentrate, one can determine the operating conditions necessary, such as the inlet dryer temperature when the dryer outlet temperature and product moisture content are targeted. This is the ratio of the energy needed to remove water from a concentrate to the energy to evaporate the same amount of pure water.

However, how could one determine this ratio, which must vary if the solids are of different composition and biological nature? A specific device was thus designed to obtain this ratio. The novel device used a precise microbalance and a desorption cell that permitted measurement of the mass change of the concentrate sample and the relative humidity of the air at the same time.

Different materials (water, skim milk, infant formulae, etc.) have been tested using this new method. The results obtained with direct (microbalance) and indirect (thermo-hygrometer) measurements were found to be highly consistent (coefficient of determination as 1). This desorption method used an airtight stainless steel cylindrical cell (Figure 4.1.1) to dry a milk concentrate sample (160 ± 1 mg) contained in a small plastic cup at 45°C. The desorption cell was filled with 120 ± 1 mL zeolite particles (Zhu *et al.*, 2011a). Water then transferred from the concentrate to the zeolites by means of the pressure gradient between them. Changes in the relative humidity of the air inside the cell during desorption were continuously monitored using a relative humidity sensor placed close to the surface of the milk sample. The change in relative humidity of the air between the concentrate and the zeolites could be measured by a thermo-hygrometric sensor connected to a computer, and the drying rate of the concentrate could be calculated from the desorption curve using the SD2P® software. By integrating several parameters of the dryer, the concentrate, and the final product requirements (e.g., evaporation capacity, airflow rates and humidity,

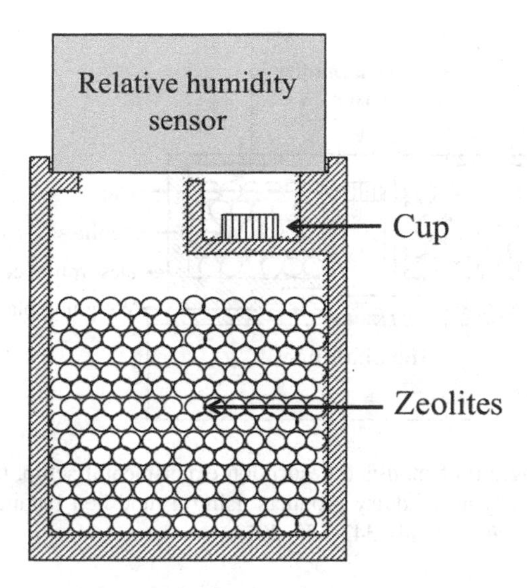

FIGURE 4.1.1 Diagram of original desorption experimental setup. (Modified from Zhu *et al.*, Prediction of drying of dairy products using a modified balance-based desorption method. *Dairy Sci. Technol.* 93(4), 347–355, 2013.)

total solid contents and temperature of the concentrate, water content of the powder in relation to water activity, energy cost, etc.), the main operational parameters of spray drying, such as inlet and outlet air temperatures, concentrate and powder flow rates, specific energy consumption, yield of the dryer, and cost (per kilogram of water evaporated or per kilogram of powder produced) could then be predicted from the mass and energy balance. Validation tests of over 30 different food concentrates have been performed by using different spray dryers (evaporation capacity from 5 $kg \cdot h^{-1}$ to 6 $T \cdot h^{-1}$), and a good match between measured and predicted parameters (±1–5% error) has been reported (Schuck *et al.*, 2009).

In this method, the drying rate is evaluated indirectly from the change in the relative humidity by considering that the area below the relative humidity curve is representative of the quantity of evaporated water. However, the relative humidity values measured at the final stage of desorption are very low and close to the detection threshold of the thermo-hygrometric sensor. This could be a source of error for prediction by the SD2P® software. Zhu *et al.* (2013) modified the device to make the measurement more directly (Figure 4.1.2).

In any case, a typical result is shown in Figure 4.1.3. Since the inner gas space is known for the devices, one can work out the rate of drying based on the *RH* change over time. This figure shows that there are at least two rate stages, as is expected for this sort of material, such as milk. The average rates can be worked out to be used for spray dryer mass balancing and energy balancing calculations. If it was used for pure water, Figure 4.1.3 would change to a kind of more constant *RH* over time. The $RH_{cumulated}$ (area under the *RH*–time curve) in the pure water evaporation experiment divided by the $RH_{cumulated}$ in the sample drying experiment (which should be >1) may indicate how much energy is needed to remove the same amount of water from the

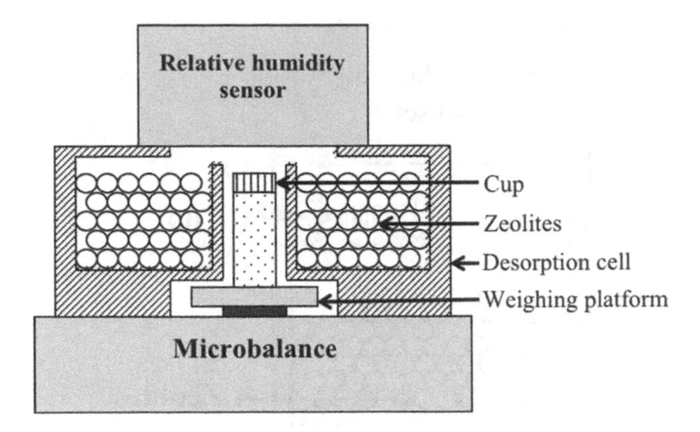

FIGURE 4.1.2 Diagram of modified desorption experimental setup. (Modified from Zhu *et al.*, Prediction of drying of dairy products using a modified balance-based desorption method. *Dairy Sci. Technol.* 93(4), 347–355, 2013.)

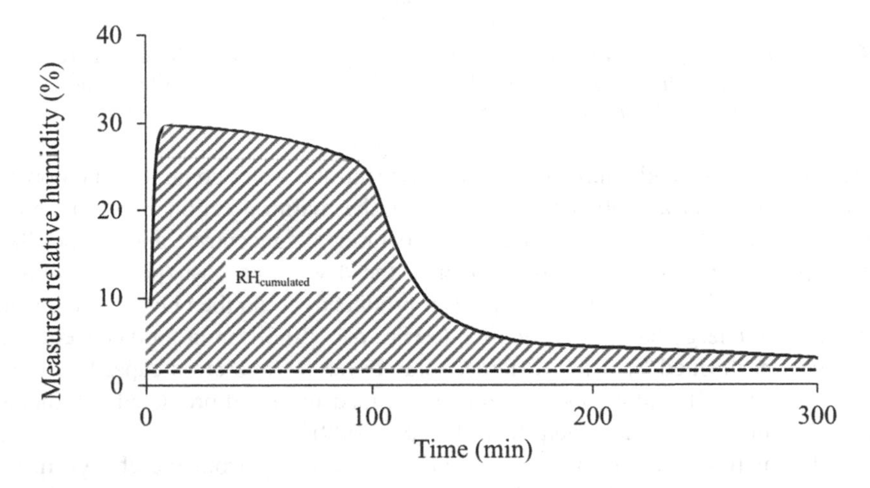

FIGURE 4.1.3 An example of the desorption curve. -: measured relative humidity of 40% w/w total solids skim milk; ---: bulk relative humidity of zeolite; $RH_{cumulated}$: cumulated relative humidity. (After Zhu *et al.*, Prediction of drying of dairy products using a modified balance-based desorption method. *Dairy Sci. Technol.* 93(4), 347–355, 2013.)

sample. This ratio may be assumed to be similar for a large dryer. The test results obtained using the device in Figure 4.1.2 are shown in Figure 4.1.4.

The work carried out by Zhu *et al.* (2013) has demonstrated an improved device for the desorption method. It has shown that the device in Figure 4.1.1 has also given consistent results that have nevertheless been confirmed to be excellent in correlating numerous plant operation data (Schuck *et al.*, 1998, 2009).

4.1.3 Effective Rate Approach (ERA)

For any spray dryer, it is always desirable to have a reliable tool that can quickly predict particle product properties based on the inlet conditions. If available, the tool

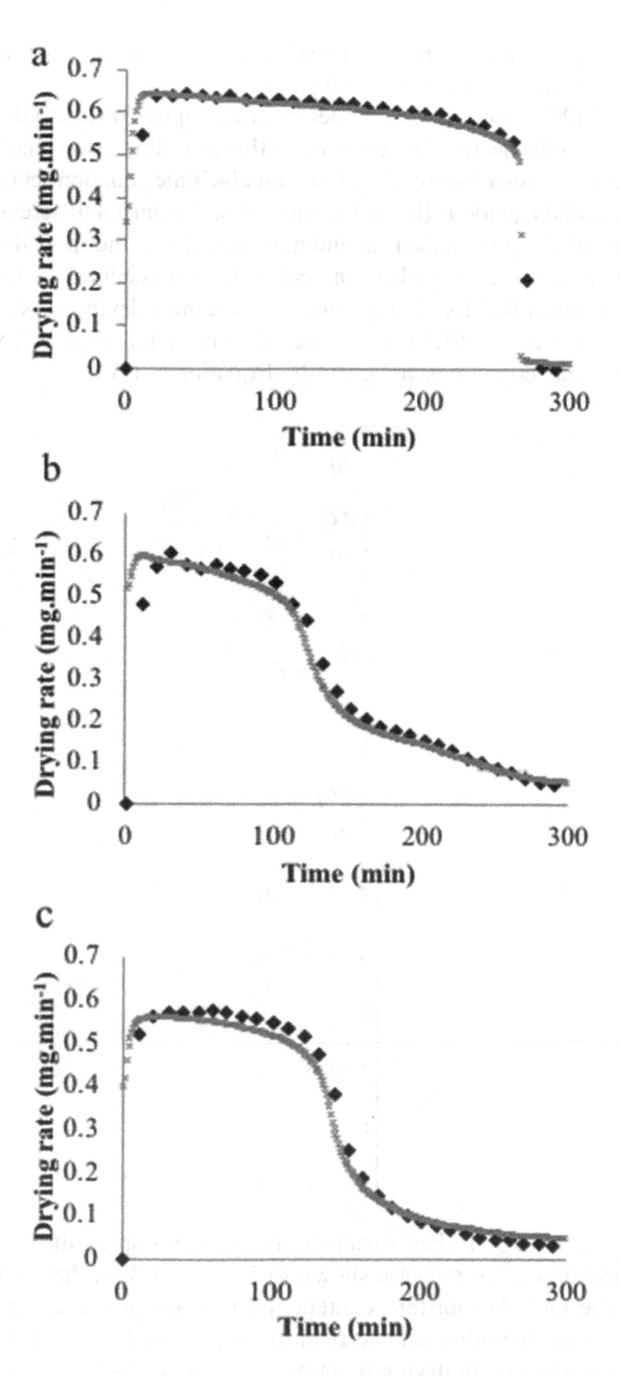

FIGURE 4.1.4 Drying rate profiles obtained during drying by desorption of distilled water. (a) 40% w/w total solids skim milk; (b) 40% w/w total solids infant formula; (c). ◆: by microbalance, ×: by relative humidity sensor. (After Zhu *et al.*, Prediction of drying of dairy products using a modified balance-based desorption method. *Dairy Sci. Technol.* 93(4), 347–355, 2013.)

can contribute significantly to powder product quality control, dryer design, operation optimization, and on-site problem shooting.

George *et al.* (2015) introduced an effective rate approach (ERA) that can realize the above-mentioned objective. By resorting to this zero dimension method, the outlet conditions can be obtained by simply solving an algebraic equation set rather than the ordinary differential equations (i.e., a 1-D model) or the partial differential equations (i.e., a 3-D model) that govern the heat and mass transfer in the spray drying process.

The following are some key ideas and calculation procedures. A set of ordinary differential equations (ODEs) that govern the dynamic drying process should be developed first. Equation (4.1.3.1) is a general form of the ODE set, whose initial conditions (i.e., inlet conditions) are given by Equation (4.1.3.2):

$$\begin{cases} \frac{dx_1}{dt} = f_1 \\ \frac{dx_2}{dt} = f_2 \\ \vdots \\ \frac{dx_i}{dt} = f_i \\ \vdots \\ \frac{dx_N}{dt} = f_N \end{cases} \tag{4.1.3.1}$$

$$\begin{cases} x_1(t_0) = x_1^* \\ x_2(t_0) = x_2^* \\ \vdots \\ x_i(t_0) = x_i^* \\ \vdots \\ x_N(t_0) = x_N^* \end{cases} \tag{4.1.3.2}$$

where x_1, x_2, …, and x_N are key variables that define a spray drying process in a particular spray dryer. For the case shown in Figure 4.1.3.1a, the six variables are, respectively, the particle moisture content, particle temperature, air temperature, particle velocity, air humidity, and vertical distance travelled by a particle. Without solving the ODEs to obtain drying dynamics over time ($t \in [t_0, t_e]$), the key idea of ERA is to use the concept of "effective rates" to develop a set of coupled algebraic equations that can be solved to obtain the outlet conditions.

$$x_i(t_e) = x_i^* + \overline{f_i} \times (t_e - t_0) \quad i = 1, 2, \ldots, N \tag{4.1.3.3}$$

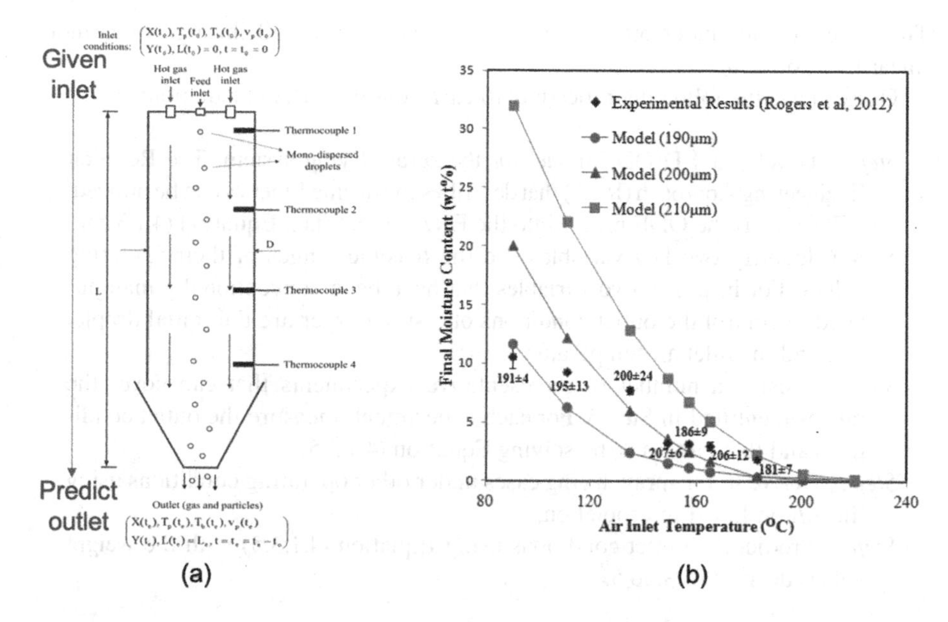

FIGURE 4.1.3.1 Application of the ERA approach to the investigation of a mono-disperse droplet spray dryer: (a) inlet conditions and outlet variables; (b) ERA predictions on particles' final moisture content for different air inlet temperatures and initial droplet diameters compared with experimental measurements. Initial droplet diameters used in the experiments are shown with the labels. (Adapted from George, O.A., Chen X.D., Xiao, J., Woo, M., Che, L., An effective rate approach to modeling single-stage spray drying. *AIChE J.* 61(12), 4140–4151, 2015.)

where $\overline{f_i}$ is the *effective rate* of evolution of the *i*th variable x_i. It can be written as a weighted mean of the initial and final values of f_i, i.e.,:

$$\overline{f_i} = w_i \times f_i\left(t_0\right) + \left(1 - w_i\right) \times f_i\left(t_e\right) \quad i = 1,\ 2, \ldots, N \tag{4.1.3.4}$$

where w_i is the weight, whose value depends on the shape of the f_i versus t curve. Thus, the ERA model becomes:

$$x_i\left(t_e\right) = x_i^* + \left(w_i \times f_i\left(t_0\right) + \left(1 - w_i\right) \times f_i\left(t_e\right)\right) \times \left(t_e - t_0\right) \quad i = 1,\ 2, \ldots, N \tag{4.1.3.5}$$

with

$$f_i\left(t_0\right) = F_i\left(x_1^*, x_2^*, \cdots, x_N^*, t_0\right) \tag{4.1.3.6}$$

$$f_i\left(t_e\right) = F_i\left(x_1\left(t_e\right), x_2\left(t_e\right), \cdots, x_N\left(t_e\right), t_e\right) \tag{4.1.3.7}$$

To obtain the effective rates, we can first calculate w_i by solving Equation (4.1.3.5) with outlet conditions measured in experiments under pre-specified initial conditions.

Then, the weights under other conditions can be determined through linear/non-linear interpolation.

One can use the following procedure to carry out the ERA calculation:

Step 1. Develop a 1-D ODE model for the spray drying system. The Reaction Engineering Approach (REA) that describes the drying kinetics can be utilized.
Step 2. Convert the ODE model into the ERA model (i.e., Equation (4.1.3.5)).
Step 3. Identify two key variables and the feasible ranges of their operating values. For instance, two variables that have been conventionally manipulated to control the outlet conditions of a spray dryer are the initial droplet size and the inlet air temperature.
Step 4. Design a number of representative experiments that can cover the ranges identified in Step 3. For each experiment, measure the outlet conditions and then obtain w_i by solving Equation (4.1.3.5).
Step 5. Derive w_i for spray drying cases under other operating conditions using linear/non-linear interpolation.
Step 6. Predict the outlet conditions using Equation (4.1.3.5) with the weight values derived in Step 5.

The capability of this approach has been demonstrated by running a mono-disperse droplet spray dryer. This dryer is an ideal experimental platform for model validation, since the inlet condition of the spray can be accurately controlled in terms of the initial droplet diameter and location, which may not be feasible in many other spray drying devices. For 20% wt skim milk, the prediction values of the final moisture content under different combinations of the inlet air temperature and the inlet droplet diameter are plotted in Figure 4.1.3.1b. It can be concluded that the prediction performance of the current ERA model is quite satisfactory, not lower than that of a complex CFD model (see CFD results in Yang *et al.*, 2015).

For most cases, it is sufficient to simultaneously manipulate two key variables for the control and optimization of a spray dryer. How to uniquely identify effective rates when more than two operating variables need to be considered simultaneously is still an open question. Furthermore, the effectiveness of the ERA concept for industrial-scale spray dryers has to be tested. Nevertheless, we believe ERA is a new and promising tool that can quickly predict dryer outlet conditions with negligible computational demand. It will be a powerful tool for process control and optimization in those industries that involve spray drying operations.

4.2 ONE-DIMENSIONAL MODELS

4.2.1 Co-Current Model

The premise of the one-dimensional model is in the tracking of droplets moving in one direction of the spray drying chamber along the length of the chamber. With the one-dimensional model, the predictive tool will be able to provide the drying history of the droplet as well as the thermal and humidity changes in the drying air along the length of the chamber, and then provide an indication of the length and size required for the intended spray drying operation. Knowing the evolution of the droplet drying history will also allow the prediction or development of particle stickiness or product quality change throughout the drying process.

This modeling framework so far assumes a plug flow movement of air and atomized droplets within the chamber (Patel *et al.*, 2010; Truong *et al.*, 2005). This is not true of most industrial dryers, in which the flow, upon reaching the bottom conical region, will tend to recirculate upwards. In essence, only the downward motion of the airflow and droplets will need to be modeled, as most of the significant portion of the drying process only occurs in the initial contact with the hot air. A key assumption here is that the atomized droplets are radially homogenously mixed and a representative initial droplet size is assumed. A graphical representation of this theoretical framework is given in Figure 4.2.1. The one-dimensional predictive framework involves the solution of a set of five equations below:

Particle momentum:

$$\frac{dv_p}{dt} = C_D \frac{18\,\mu_b\, Re}{24\,\rho_p\, d_p^2}\left(v_b - v_p\right) + \frac{g}{\rho_p}\left(\rho_p - \rho_b\right) \tag{4.2.1}$$

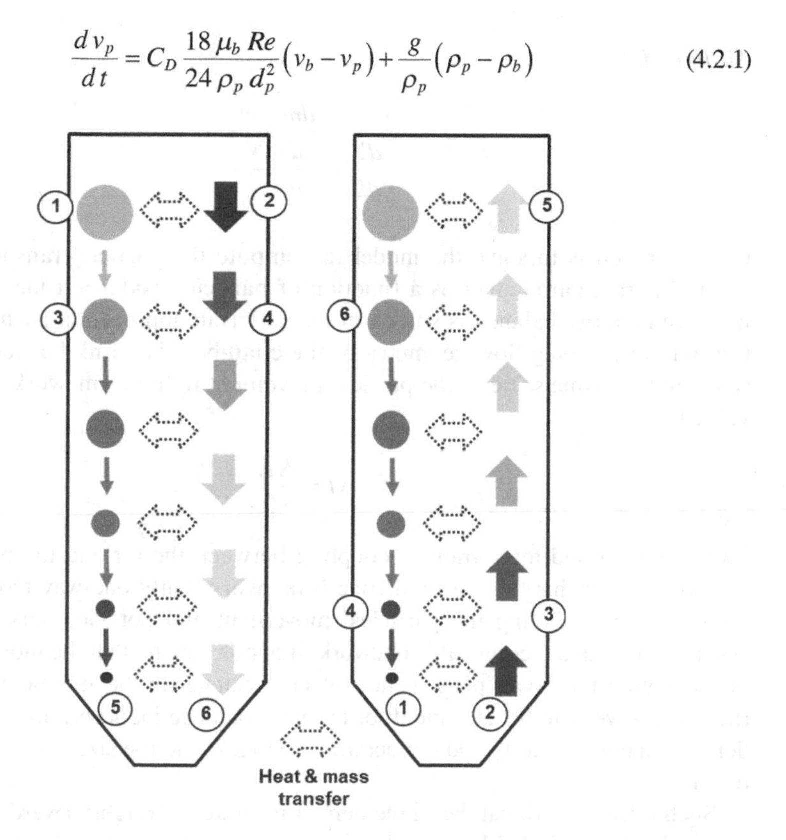

FIGURE 4.2.1 Comparison of the complexity in the two-way coupling for the one-dimensional predictive framework. Left: co-current spray dryer; right: counter-current spray dryer. (From Woo, M.W., Advances in production of food powders by spray drying, in *Advanced Drying Technologies for Food*, edited by Mujumdar, A.S., Xiao, H.W. Taylor & Francis, Boca Raton, 2019. With permission.)

Particle moisture content:

$$\frac{dm_w}{dt} = h_m A_p \left(\psi \rho_{v,\text{surface}} - \rho_{v,\text{bulk}}\right) \tag{4.2.2}$$

Particle temperature:

$$\frac{dT_p}{dt} = h A_p \left(T_b - T_p\right) - \frac{dm_w}{dt} \Delta H_L \tag{4.2.3}$$

Air temperature:

$$\frac{dT_b}{dL} = \frac{\left(\frac{dm_w}{dt}\frac{\theta}{v_p}\right)\left[\Delta H_L - C_{p,v}\left(T_b - T_p\right)\right] - \frac{\theta}{v_p} h A_p \left(T_b - T_p\right) - U\left(\pi D\right)\left(T_b - T_{amb}\right) - \dot{V} \rho_b \Delta H_L \frac{dY}{dL}}{\dot{V} \rho_b C_{p,b}} \tag{4.2.4}$$

Air humidity:

$$\frac{dY}{dL} = \frac{\frac{dm_w}{dt}\frac{\theta}{v_p}}{\dot{m}_{b,\text{dry}}} \tag{4.2.5}$$

One approach is to solve the model to compute the particle transport equations in the Lagrangian manner as a function of particle residence time, while the air mass and energy balance is solved in the Eulerian approach at each spatial location within the plug flow geometry of the chamber. The link between the length scale and the time scale of the particle movement in the framework is the particle velocity.

$$\Delta t = \frac{\Delta L}{v_p^n} \tag{4.2.6}$$

Two-way heat and mass transfer coupling between the air and the powder is also incorporated in this type of predictive framework. Only one-way momentum coupling (the air affecting the particles' movement and not vice versa) is typically assumed in such a theoretical framework. It is noteworthy that the momentum transport equation for the air phase is not solved. Changes in the air velocity throughout the spray tower can be accounted for by mass balance incorporating changes in the density (due to humidity and temperature changes) and the size and geometry of the tower.

Such a framework can be implemented in quite a straightforward manner using spreadsheet tools (e.g., Microsoft Excel) or a more advanced mathematical tool such as MATLAB®. This predictive framework will be able to provide a quick estimation of the spray dryer's performance and can even be used as a training tool for operators.

4.2.2 Counter-Current Model

Application of the one-dimensional model to counter-current tower modeling mainly involves the same equations listed in the preceding section. The main challenge is then in numerically implementing the solution. The equations pertaining to the droplets need to be solved in the direction opposite to the solution of the air phase. Such complexity is illustrated in Figure 4.2.1. In the counter-current configuration, the heat and mass transfer between the particle and air at positions 1 and 2 will propagate to progressively cooler air conditions at position 3. The slightly cooler air at position 3 will subsequently undergo heat and mass transfer with the particle at position 4 (which numerically is the particle at position 1 in the preceding time step) forming a non-linear numerical loop between each discretized "Eulerian" or "Lagrangian" step size. Such loops are interconnected and will further propagate across the whole system (e.g., positions 5 and 6).

In view of the numerical loops described, while the co-current model can be solved directly in a spreadsheet manner by the "dragging" solution, propagating it forward in time or chamber space, iterations are required for the counter-current framework. From the authors' experience, when implementing different numerical schemes in Microsoft Excel-based spreadsheets with the aim of developing a quick and easy predictive tool, the speed and the numerical stability of different iterative loops will need to be considered (Razmi *et al.*, 2019). Drawing inspiration from the co-current approach, one may use the built-in automated iterative circular looping feature in Microsoft Excel to develop a "dragging" marching forward method for the computation (Figure 4.2.2). While this approach does not require coding nor implementation of any macros for iteration, due to the fine step size and the high total number of discretized steps this approach is highly unstable. With very large datasets (a high total number of discretized steps), the built-in circular looping feature also takes a very long time to compute. A more stable approach is to use the numerical scheme shown in (Figure 4.2.3). This is similar to typical Eulerian–Lagrangian iterative schemes in CFD simulations. Implementing this in the Microsoft Excel framework will require the use of macros and Visual Basic capabilities.

The counter-current framework will also require capturing the possible entrainment point for the powder, a feature not required to be captured in the co-current framework. In the report by Ali *et al.* (2014), the possible entrainment of particles is neglected by enforcing the terminal velocity as the lower particle velocity limit. Such an assumption may have a non-physical limitation assumption if the simulation is "pushed" to the limits of counter-current spray drying, as it implies that excessively high airflow rates can be used without any operation limitation imposed by the possible entrainment phenomenon. In addition to being an important physical phenomenon, the reversal also has numerical significance to avoid the zero or negative particle velocity condition, which will lead to numerical "blow-out"; this is because, in the current recommended framework, the particle velocity is used to link the Eulerian and Lagrangian simulation frameworks. Precisely pinpointing the entrainment point is not trivial, because

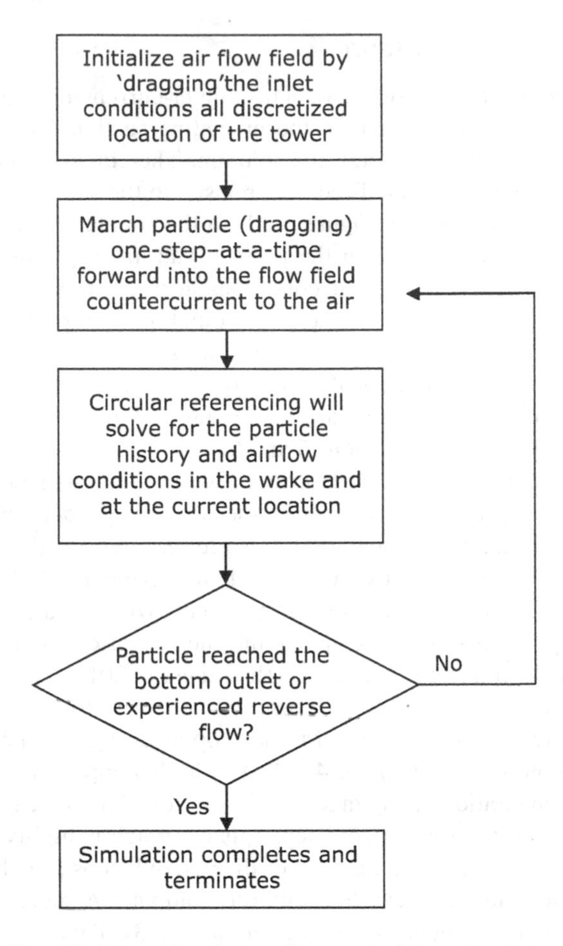

FIGURE 4.2.2 Unstable "marching forward" numerical loop for the counter-current predictive framework. (Adapted from Razmi *et al.*, What is important in the design of counter current spray drying towers? *Proceedings of CHEMECA*, September 29–October 2, 2019, Sydney, Australia, 2019.)

as the particle velocity approaches zero, the time step size or integration time scale becomes progressively large. The authors have developed a set of numerical criteria for spray tower configurations (not fountain-like configurations) to overcome this limitation with the computation of a particle terminal velocity. In the following theoretical framework, the particle terminal velocity for the latest time step will be computed using the air velocity and properties at the previous particle time step.

$$v_{p,\text{terminal}} = v_b - \frac{g}{\rho_p}\left(\rho_b - \rho_p\right)\frac{24\,\rho_p\,d_p^2}{C_D 18\,\mu_b\,Re} \tag{4.2.7}$$

On the basis that the droplets are injected downwards, the following four situations can then be encountered in the simulation:

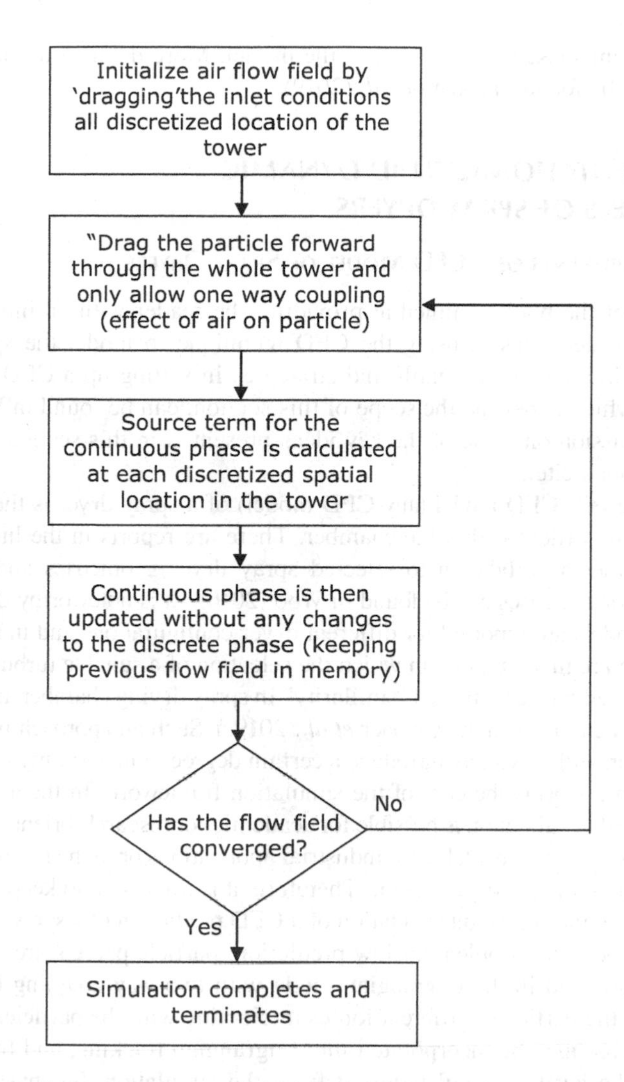

FIGURE 4.2.3 Stable iterative numerical loop for the counter-current predictive framework. (Adapted from Razmi *et al.*, What is important in the design of counter current spray drying towers? *Proceedings of CHEMECA*, September 29–October 2, 2019, Sydney, Australia, 2019.)

Scenario 1.	[Downward particle velocity] + [Upward terminal velocity]
Scenario 2.	[Upward particle velocity] + [Upward terminal velocity]
Scenario 3.	[Downward particle velocity < downward terminal velocity] + [Downward terminal velocity]
Scenario 4.	[Upward particle velocity] + [Downward terminal velocity]

For Scenarios 3 and 4, the particle velocity will then adopt the downward terminal velocity. In Scenario 1, the integration of the particle trajectory can be continued as usual. In Scenario 2, the particle is deemed to have reached the reversal point and is removed from subsequent sections of the model. More details on this numerical technique can be found in Razmi *et al.* (2019).

4.3 COMPUTATIONAL FLUID DYNAMIC MODELS OF SPRAY DRYERS

4.3.1 Components of a CFD Model of Spray Dryers

This section of the book is aimed at providing the reader with an introduction to some key considerations in using the CFD technique to model the spray drying process. Specific numerical details and strategies in setting up a CFD model of a spray dryer, which is beyond the scope of this section, can be found in Woo (2016). Detailed discussion on some of the key ideas presented in this section can also be found in the book cited.

At the core of a CFD model (any CFD model) of a spray dryer is the simulation of the turbulent airflow within the chamber. There are reports in the literature providing experimental validation to selected spray dryer geometries and operations (more details on the cases can be found in Woo (2016)). It is noteworthy that different turbulence models are reported for different dryer configurations and in many cases. As such, there are many reports in which the selection of a suitable turbulence model was mainly based on past reported "similarity" in spray drying chamber and operation (discussed in the introduction of Jubaer *et al.,* 2019c). Such an approach in turbulence model selection will inevitably introduce a certain degree of uncertainty in the airflow prediction, which forms the core of the simulation framework. In the authors' opinion, while detailed validation is possible for academic- or research-oriented simulation work, such luxury is not available for industrial applications or in the routine use of the CFD technique for spray dryer design. Therefore, it is important to keep in mind this aspect of uncertainty in the interpretation of a CFD prediction of a spray dryer.

Building upon the turbulent airflow prediction, particle parcels are then tracked within the flow field in the Lagrangian mode with two-way coupling between the air phase and the particles. Different forces interacting with the particles in addition to the drag force may be incorporated into Lagrangian tracking, and this certainly depends on the level of detail required from the simulation. Stochastic turbulent dispersion of the particles is typically accounted for in most reported CFD simulations. Incorporated into the Lagrangian particle tracking framework are all the other important sub-models, which "convert" the entire CFD simulation framework into a spray dryer CFD simulation. Some of these sub-models include:

1. Droplet drying model
2. Particle–wall interaction model
3. Agglomeration model

While the droplet drying model has already been discussed in Chapter 2 of this book, the particle–wall interaction model and the agglomeration model will be discussed later on in this chapter. Regardless of the type of sub-model, particle tracking model and turbulent airflow prediction used, underpinning the entire simulation framework is another aspect that has to be carefully considered: to undertake the simulation in the steady state or transient state? This will be discussed in greater detail in the next section.

4.3.2 The Trouble with Transient Simulations

This section will center around the Euler–Lagrangian CFD simulation framework of the spray drying process. Another basis adopted here is that such simulation typically utilizes two-way coupling for the development of the flow field. Details on this aspect of a CFD simulation of spray dryers can be found in other publications (Woo, 2016). Figure 4.3.1 shows two different numerical approaches in which a steady and a transient spray dryer simulation can be undertaken. Solutions with the transient framework will certainly involve higher computational requirements and longer computation time. The key differences are in the simulation time required for the development of the flow field incorporating particle injection.

In the steady state framework, each particle is introduced and tracked through the flow field one at a time. The numerical iterations to incorporate the two-way momentum, heat, and mass coupling is "constant" at each time step and is typically determined by the number of particle injections pre-set to represent the initial droplet size distribution injected into the simulation domain. The memory requirement is also lower, as the solver has to track only one particle at a time and the source terms are accumulated across all the particles injected subsequently. In the steady state framework, the particle time integration step size does not affect the numerical requirement in the two-way coupling development of the flow field.

In contrast, within the transient framework, particles are continuously injected at a predetermined injection rate and at each time step, they are only progressed "one step" within the domain. Therefore, the number of particles in the simulation domain will be progressively increased, which further increases the number of particle tracking routines required and the memory for each particle in the simulation domain. For the proper development of the flow field, one then has to continuously allow a "steady" number of particles to accumulate within the system before any interpretation of the simulation can be undertaken. This total of "steady" particle numbers is a function of the injection frequency and, most importantly, determined by the size of the simulation domain. From the authors' experience, attempts to reduce the injection frequency with the intention of reducing the total number of injected particles, particularly for large drying chambers, may lead to different chamber outlet predictions and is not recommended. All these different factors will significantly extend the simulation time when compared with the steady state simulation. For large industrial-scale simulations, the flow field may take several days to develop with typical desktop computing (not network) resources. It is also noteworthy that transient simulations may require different post-processing of the results. Due to the transient nature of the flow field and transient particle tracking, some reports suggested the use of time-averaged sampling of the particles (Jin and Chen, 2009a, 2010). We have evaluated the use of a pseudo-steady approach to post-processing of the particle tracking results, to avoid the long time-averaging sampling approach. More details can be found in Woo *et al.* (2011b).

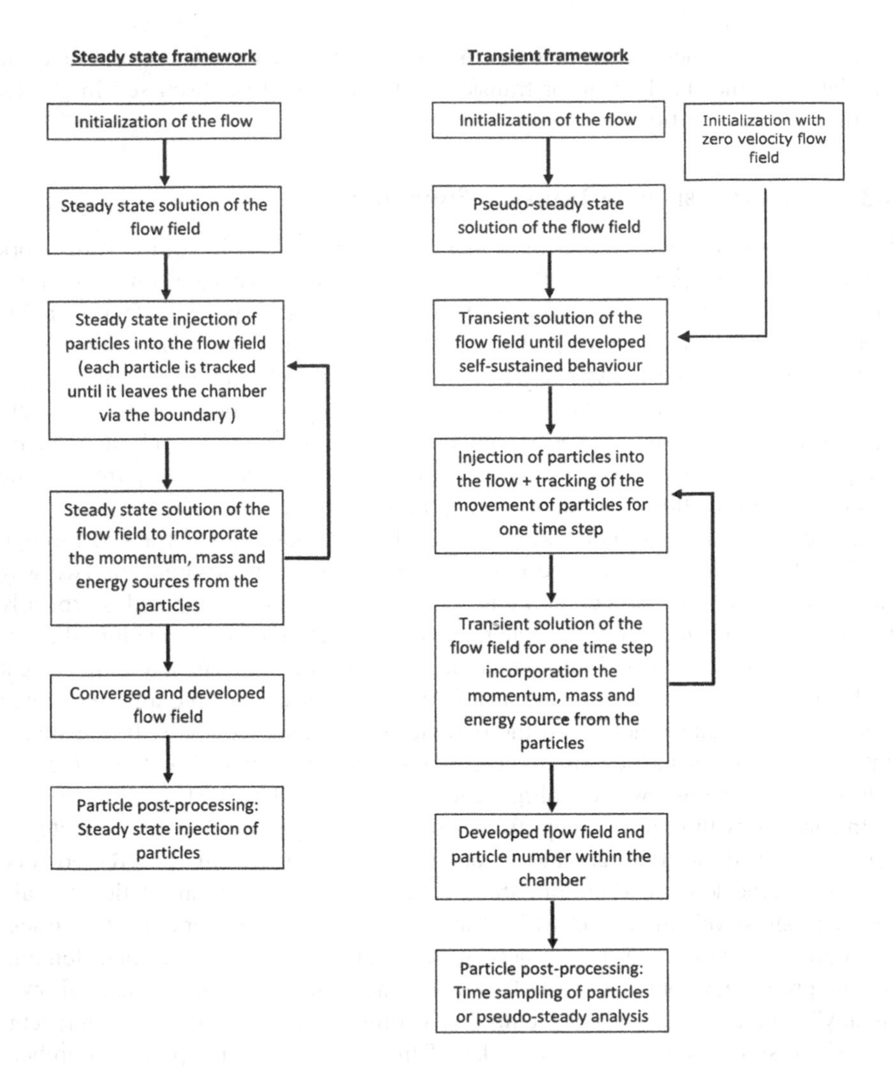

FIGURE 4.3.1 Numerical approaches to undertaking steady or transient CFD simulations of spray dryers. (Adapted from Afshar, S., Jubaer, H., Chen, B., Xiao, J., Chen, X.D., Woo, M.W., Computational fluid dynamics simulation of spray dryers: transient or steady state simulation? *Proceedings of the 21st International Drying Symposium*, September 11–14, 2018, Valencia, Spain, 2018.)

In view of the resource challenges to transient simulations, the next question is: how do we choose between steady state and transient simulations? This will be an important question if the CFD simulation technique is to be used in a more routine manner for industrial applications. A review of the reported simulations in the literature reveals a contradictory picture (Table 4.3.1). Earlier reports mostly employed the steady state approach with two-dimensional simulation domains. The use of the two-dimensional simulation domain may have the tendency to numerically suppress the possibility of the fluctuation behavior, which in essence is a

TABLE 4.3.1
Compilation of Steady and Transient CFD Simulations of Spray Dryers

Year	References	Steady or transient	Chamber geometry	Focus
1993	Oakley and Bahu (1993)	Steady	Co-current cylinder on cone	Airflow and droplet drying study (2-D model)
1997	Kieviet *et al.* (1997)	Steady	Co-current cylinder on cone	Airflow pattern study (2-D model)
1998	Frydman *et al.* (1998)	Steady	Co-current tower	Superheated steam spray drying (2-D model)
1999	Straatsma *et al.* (1999)	Steady	Co-current cylinder on cone	Airflow and droplet drying (2-D model)
2002	Harvie *et al.* (2002)	Steady	Co-current tower	Droplet drying and deposition modeling
2004	Verdumen *et al.* (2004)	Steady	Co-current cylinder on cone	Agglomeration modeling
2004	Huang *et al.* (2004)	Steady	Co-current cylinder on cone	Turbulence model evaluation
2004	Langrish *et al.* (2004)	Transient	Co-current cylinder on cone	Investigation on the influence of inlet swirl
2007	Kota and Langrish (2007)	Transient	Co-current cylinder on cone	One-way pseudo-steady particle analysis
2007	Blei and Sommerfeld (2007)	Steady	Co-current cylinder on cone	Airflow and particle movement
2008	Woo *et al.* (2008b)	Steady	Co-current cylinder on cone	Droplet drying modeling (2-D axisymmetric)
2009a	Jin and Chen (2009a)	Transient	Co-current cylinder on cone	Study on the effect of different particle size on spray dryers
2009b	Jin and Chen (2009a)	Transient	Co-current cylinder on cone	Gas–particle interaction study
2009	Woo *et al.* (2009)	Transient	Co-current cylinder on cone	Airflow stability study
2009	Mezhericher *et al.* (2009)	Steady	Co-current cylinder on cone	Droplet drying modeling
2009	Fletcher and Langrish (2009)	Transient	Co-current cylinder on cone	Evaluation of SST turbulence model for spray dryers
2010	Ullum *et al.* (2010)	Steady	Co-current cylinder on cone	Wall deposition study
2010	Anandharamakrishnan *et al.* (2010)	Steady	Co-current cylinder on cone	Investigation on particle histories in spray drying
2010	Jin and Chen (2010)	Transient	Co-current cylinder on cone	Wall deposition study
2010	Gabites *et al.* (2010)	Transient	Co-current with bottom static bed	Coherent transient behavior study
2011	Woo *et al.* (2011)	Transient	Co-current tower	Low-velocity airflow drying
2011	Jin and Chen (2011)	Transient	Co-current cylinder on cone	Entropy generation in spray dryers
2011	Woo *et al.* (2011)	Transient	Co-current tower	Low-velocity droplet crystallization analysis

(*Continued*)

TABLE 4.3.1 (CONTINUED)
Compilation of Steady and Transient CFD Simulations of Spray Dryers

Year	References	Steady or transient	Chamber geometry	Focus
2012	Mezhericher *et al.* (2012)	Transient	Co-current cylinder on cone	Droplet collision modeling
2012	Woo *et al.* (2012a)	Transient	Co-current cylinder on cone	Transient flow and wall deposition study
2012	Wawrzyniak *et al.* (2012)	Transient	Counter-current tower	Airflow pattern study
2013	Jongsma *et al.* (2013)	Transient	Co-current cylinder on cone	Large-scale eddy simulation study
2015	Jaskulski *et al.* (2015)	Steady	Counter-current tower	Agglomeration modeling
2015	Mezhericher *et al.* (2015)	Transient	Co-current cylinder on cone	Droplet drying modeling
2015	Malafronte *et al.* (2015)	Steady	Co-current cylinder on cone	Coalescence and agglomeration modeling
2015	Yang *et al.* (2015)	Transient	Co-current mono-disperse atomizer	Mono-disperse injection pattern
2016	Schmitz-Schug *et al.* (2016)	Steady	Co-current cylinder on cone	Modeling lysine loss in dairy products
2017	Ali *et al.* (2017)	Steady	Counter-current tower	Particle–wall collision study
2017	Jaskulski *et al.* (2017)	Steady	Co-current tower	Protein denaturation modeling
2017	Wawrzyniak *et al.* (2017)	Transient	Counter-current tower	Air–particle flow behavior
2017	Jubaer *et al.* (2017)	Transient	Co-current tower	Evaluation of different shrinkage models

three-dimensional phenomenon (Fletcher and Langrish, 2006). Later reports with three-dimensional simulation domains constitute a mixture of steady state and transient simulation approaches (Table 4.3.1). For future workers in this area, to ascertain if the simulation is steady or transient, it will be important to check if the steady state simulations are truly steady or are actually changing, albeit slightly, at subsequent iterations, particularly in regions along the edge and at the end of the central air jet. If such situations are observed, this may be an indication that a transient simulation is required. In any case, from the authors' experience, significant flow field flipping may be observed in steady state simulations of highly swirling flows in spray drying chambers, which will be a clear indication of the need for transient simulations.

4.4 BRIDGING THE GAP BETWEEN SPRAY DRYER MODELING AND OPERATION

In this section of the book, we will discuss two main challenges when implementing dryer modeling for industrial application. The first challenge concerns the level of accuracy, particularly in the prediction of the product moisture content, which can be expected from black box, one-dimensional, or CFD models of the spray drying process. In general, in the authors' opinion, the accuracy may be in the range (or order) of +1%–2% wt moisture content if set up correctly (or maybe an even larger range but in the same order). Therefore, at the moment, the models developed will be more suited to guide operators and designers on how the dryer will behave with specific changes in the operating conditions. It will be useful for troubleshooting and to determine suitable operating ranges, particularly when trying out new formulations. It is not suitable as a routine quality control tool that requires the control of the moisture content to an accuracy in the order of 0.1% wt.

Many factors contribute to these current limitations in accuracy. For the black box and the one-dimensional model, one major factor is the simplicity of the approach, which does not account for many of the smaller details in spray drying design and operation. For the more detailed CFD simulations, the complex and non-linear behavior of the modeling framework will have some inherent discrepancy. The selection of certain sub-models within the predictive framework, e.g., the airflow turbulence models for such a complex system, is still quite arbitrary. Discrepancies may also be introduced by difficulty in obtaining reliable boundary conditions for the model.

One common difficulty faced is in determining the initial droplet size from the atomizer. A sensitivity study revealed that even a variation of 20 microns in droplet size is sufficient to lead to significant change in dryer operation prediction. Most atomizers (nozzles) are rated and characterized by the vendor using water. This will be significantly different from most concentrated materials sprayed in the industry. Therefore, the industry will have limited knowledge of the actual droplet size distribution from the nozzles, which is required as an input to the predictive framework. If measurements are not available for the setting up of the models, this parameter will need to be approximated either from rough back calculations from the product or estimated based on vendor specifications.

In planning a spray dryer modeling project, from the authors' experience, if the material properties and drying kinetics such as those described in Chapter 2 are not available, a significant amount of resources and time will be required to measure them. The resources required may be comparable to those required to set up and numerically solve the models. This needs to be kept in mind when planning and budgeting for a modeling endeavor for the spray drying process.

4.5 MODELING STICKINESS

4.5.1 Why Model Stickiness?

Mathematically capturing the stickiness of the particles or droplets in spray dryer simulations is important as the basis for particle–wall deposition modeling and agglomeration modeling. While particle–wall deposition modeling is important in determining the yield from the process, agglomeration modeling will be important in the prediction of the functionality of the powder produced.

4.5.2 Different Models and Their Comparison

One main approach to predicting stickiness is to determine whether or not the stickiness of the particle is independent of the dynamics of the powder associated with its movement within the chamber. The prediction of the sticky state is also quite complex, because it is a combination of the degree of wetness of the powder as well as the temperature of the powder. Higher moisture content and higher powder temperature will lead to more stickiness and vice versa. The glass transition phenomenon can be used to delineate how these two factors contribute to the stickiness of the powder (Adhikari *et al.*, 2004; Woo *et al.*, 2008b). The glass transition temperature between multiple components (solute and moisture) in a particle can be calculated using the Gordon Taylor equation:

$$T_g = \frac{w_1 T_{g,1} + k_{T_g} w_2 T_{g,2}}{w_1 + k w_2} \tag{4.5.1}$$

The second component in Equation (4.5.1) is normally taken as the water component. If the particle temperature is higher than the glass transition temperature by about 20°C, the powder will then be sticky. It should be noted that this threshold, in essence, is a loose criterion, with reported applicability of this threshold for sugar-based powder. This threshold is also widely accepted by the dairy industry (from the authors' experience). In some spray drying applications, however, the surface composition of the powder may be significantly different when compared with the bulk composition. This is particularly true of powder with high protein. Readers interested in gaining a better understanding of the potential deviation from the +20°C glass transition-based sticky point may refer to Hogan and O'Callaghan (2013). There is also experimental evidence to suggest that the glass transition-based sticky criterion – in fact, the stickiness of powder in general – can be affected by the dynamics of the powder colliding. Such dynamics include the effect of the collision angle and velocity on the stickiness of the powder (Zuo *et al.*, 2007).

The implementation of the glass transition-based stickiness prediction model (and the similar direct sticky point model) has been reported by several authors for the prediction of powder–wall deposition (Harvie *et al.*, 2002; Woo *et al.*, 2010). In the implementation reported, there have so far been no attempts to incorporate the collision dynamics mentioned earlier in the prediction of stickiness. This has the advantage of simplicity, and in the simplified heat and mass model or even the one-dimensional model of spray dryers, there is also no need for the incorporation of the collision dynamics. The use of such glass transition-based (or sticky point-based) cut-off criteria for stickiness prediction in the spray drying predictive framework is the potential sudden cut-off between deposit free and heavy depositing spray drying operations. From the authors' experience, this may not be realistic.

On the other end of stickiness prediction, there are approaches that do not utilize a cut-off criterion but rather a "continuous" approach to modeling the stickiness of the powder delineated by the rigidity of the particle. The EDECAD (Efficient design and control of agglomeration in spray drying machines) project proposed a viscous model approach to predict the degree of particle rigidity in agglomeration modeling (Verdumen *et al.*, 2004) in resisting the penetration of powder during collision. The approach is mainly based on the viscosity of the powder at different solute concentrations. One main challenge is in the estimation of powder viscosity at very low moisture content (when the powder is actually solid-like) and a relatively large exponential extrapolation is still required from relatively liquid-like conditions. The applicability of such a large extrapolation remains unclear. In that series of reports, the Ohnesorge number describes the relative significance of the fluid (droplet) viscosity and the inertia and surface tension (Verdumen *et al.*, 2004), differentiating droplet–droplet interactions or viscous (sticky) droplet–particle interactions.

In a similar vein, Woo *et al.* (2010) proposed a viscoelastic approach in which the rigidity of the particle is determined by the storage and loss modulus of the particle. The effect of moisture content on rigidity was incorporated using the glass transition temperature as the reference for the viscoelastic Williams–Landel–Ferry (WLF) shift factor. Both these approaches incorporate the effect of collision dynamics in determining the stickiness of the powder. This approach, however, has yet to be incorporated into large-scale CFD simulations of spray dryers. Measurement of the viscoelastic properties at different levels of moisture content and temperature may also be less straightforward when compared with the glass transition of the viscosity measurements required in other forms of the model.

While the approaches described so far account for the stickiness, in essence they are based on the rigidity of the particles (even the sticky point criterion approach). There are some reports that also incorporate van der Waals forces into the computation of the critical restitution factor in agglomeration modeling (Jaskulski *et al.*, 2015). At the moment, there is yet to be a comprehensive report evaluating the relative significance of both types of approaches in the CFD simulation domain. The most simplistic approach to model particle stickiness is to simply assume that all the particles are sticky without differentiating between the states of the particles. This approach has been reported by numerous reports with respect to the prediction of particle–wall interaction, leading to the prediction of the spray drying process yield.

Such an approach provides an upper bound in the prediction of the maximum deposition rate within the drying chamber.

In contrast to using the sticky point as the criterion, another commonly adopted approach is to use the stick-upon-contact approach; if a particle is simulated to touch the wall, it will be deemed deposited and is removed from the simulation (while releasing its moisture content as mass source into the simulation domain). Some reports in this area are Huang *et al.* (2004); Woo *et al.* (2008b); Jin and Chen (2009b).

4.5.3 Measuring Stickiness Properties and Their Challenges

Glass transition measurements can typically be undertaken using the digital scanning calorimeter. The technique involves progressively increasing (and/or decreasing) the temperature of a hermetically sealed sample and measuring the changes in the energy flow through the sample. Any changes in the structure of the sample (e.g., undergoing plasticization at the glass transition temperature) will be detected and delineated by changes in the energy flow pattern. It is noteworthy that in such measurement, the softening process occurs over a range of temperatures surrounding the glass transition temperature and the average central temperature within the range is typically used in stickiness modeling. Mechanical methods such as dynamic mechanical thermal analysis (DMTA) are typically used for glass transition measurements of solid or liquid samples. This technique involves rapid oscillating shear to the sample while its temperature is progressively increased. Changes in the storage and loss modulus indicate the softening process surrounding the glass transition point. Direct measurements of the sticking point can also be undertaken using a modified version of the viscometer. The premise of the technique is to measure the sudden changes in the shear stress as the temperature of the particle is progressively increased. A more comprehensive review of the measurement techniques available can be found in Adhikari *et al.* (2004). These techniques described above essentially measure the state of stickiness of the powder without incorporating the effect of collision dynamics. A unique particle stickiness gun approach was developed by Massey University and is capable of measuring the stickiness of particles as a function of plasticization and impacting conditions analogous to those found in spray drying chambers (Zuo *et al.*, 2007). In a similar vein, GEA™ has also developed a dynamic tack measurement approach, integrating with the single droplet acoustic levitation dryer to measure particle stickiness during the particle formation process under high-speed collision dynamics.

4.6 MODELING AGGLOMERATION

4.6.1 Why Model Agglomeration?

Forced agglomeration is used particularly in the spray drying of food powder to improve its dispersion and dissolution behavior. Agglomerated powder consists of irregularly packed partially fused powder particulates of various sizes. As a result, larger and more porous particulate structures are formed, providing enhanced dispersion and dissolution behavior to the powdered product. Due to the larger aggregates formed and the more irregular packing, the powder product will also be more flowable and easier to handle.

Such forced agglomeration is typically produced by the return of fine particulates recovered from the exhaust cyclone system from the spray drying chamber to the atomization region to be in contact with the partially dried particles (which are still sticky) from the atomized feed. Please refer to Section 3.2 of this book. Therefore, in addition to improving the powder quality as mentioned, agglomeration in spray drying also serves to make full potential of fine particulates, effectively maximizing the output from the process. The key lies in carefully controlling the return position and the drying process such that the fine particulates come into contact with the semi-dried particulates from the nozzle. Such interactions will be referred to as particulate–particulate collisions and are required to form particulate interactions.

If the interaction is not well controlled, the fines returned may predominantly collide or interact with the liquid droplets. Instead of producing a partially fused particulate structure, the resultant structure may be more of a coalesced form, leading to large particulates but not resulting in a more porous structure. Such a structure is not desired, as it does not enhance the dispersion and dissolution behavior of the product.

Apart from relying on the return of fines, any approach to generate agglomerate structures must align the nozzles such that there is an overlap in their spray cone. The degree and position of the overlap will have to be balanced with the rate of dehydration of the droplets, so that partially dried particles from the overlapping nozzles can collide and form forced agglomerates. Similarly, for such an approach, coalescence needs to be minimized to ensure that a relatively large and porous agglomerate structure is formed.

In all of these cases, it is clear that there is a need to distinguish regions within the spray drying chamber that allow predominantly sticky particle–sticky particle collision or interaction. As discussed, if the colliding particles are too wet, this will then lead to undesired coalescence. At the other end of the spectrum, if the colliding particles are too dry, the contact between the particles may be very minimal (minimal fusing), which can potentially lead to relatively weak agglomerate structures. Such structures may be easily broken during post-process handling or packing. In addition to controlling the state of "stickiness" of the powder, forced agglomeration will only significantly occur if the particulate number density is sufficiently high. Otherwise, there will be insufficient agglomeration formed.

Currently, such delicate agglomerate control is achieved in the industry by trial and error. In this chapter, we will discuss on the development of CFD-based models that aim at providing predictive capability to guide agglomeration control in the drying chamber. It is noteworthy that the aim of the models discussed subsequently are meant to model the agglomeration of entrained or free moving particulates. They are not aimed at modeling the occurrence of agglomeration on the wall, forming agglomerate layers on the wall (Francia *et al.*, 2015).

4.6.2 Rigorous Particulate Agglomeration Approach

The most fundamental approach to modeling agglomeration would be to employ DEM (Discrete Element Modeling)-like simulations in which each agglomerating primary particle is tracked before and after agglomeration formation. In other words, the structure of the agglomerate is explicitly modeled and the forces "holding" the agglomerate together are continuously tracked and stored in memory during the simulation (Sommerfeld and Stubing, 2017). In such a DEM approach, a relatively small time step is required to resolve the development of the whole collision process. While providing the most detailed approach to agglomeration modeling, such a high computational requirement may make it unsuitable, at the moment, for large-scale simulations; as such, we have not seen any large implementation of this approach. We will only focus on currently reported approaches for large-scale CFD simulation of the spray drying chamber. Readers interested in this approach are directed to the references cited above.

4.6.3 Equivalent Fused Particle Approach

In contrast to the approach above, all the reported large-scale CFD simulations of spray dryers utilize instantaneous particle–particle collision. Within such a framework, collisions will not be explicitly modeled (detailed mathematical tracking throughout the collision process); rather, once a collision is detected, the agglomeration model is used to compute an "instant" outcome from the collision. A pioneering work in this area is reported by the EDECAD project (Verdumen *et al.*, 2004). There are several other reports on the modeling of particle agglomeration employing similar modeling structures. The agglomeration modeling routine typically involves several sub-models.

The first sub-model is in the determination of the colliding particles. Although, in the CFD predictive framework, particles are discretely tracked, the subroutine used in several reports utilized the statistical sampling of fictional colliding particles. In essence, the colliding particles are not directly determined from colliding pairs of particles tracked in the simulation domain. Details of the fictional colliding particles scheme is given in the references cited here. In the authors' opinion, such a subroutine may be necessary, as the reported simulation work has hitherto been undertaken in the steady state simulation framework, where every particle injected into the simulation is tracked, one by one, until it leaves the simulation domain. There will be no situation where multiple particles are tracked simultaneously, in contrast to the transient simulation framework (Figure 4.3.1). Given that three-dimensional CFD simulations of spray dryers are prone to exhibit transient flow behavior, there is a need to further evaluate or develop a suitable subroutine to determine the colliding particles in the transient framework. For more information on the application of the steady state or transient frameworks for CFD modeling of spray dryers, refer to Section 4.3.3.

Once the colliding particles are identified, the next subroutine in the agglomeration model should determine the nature of the collision, whether droplet–droplet, droplet–particle, or particle–particle collision. The Ohnesorge number, which describes the ratio of the viscous characteristics to the surface tension characteristics of the droplet, is typically used as a simplified criterion to determine if the collision is of a droplet or a particle. Depending on the form of the collision, different criteria are then applied to determine collision outcome (Jaskulski *et al.*, 2015; Ali *et al.*, 2017).

In particular, for the particle–particle collision, the EDECAD framework further suggested the use of a viscous model, which tracks the degree of penetration of the colliding semi-solid particles into each other as part of determining the outcome of the collision.

If the colliding particulates are determined to agglomerate or coalesce, at the moment, both colliding particles are numerically fused into a single equivalent spherical particle conserving the mass and volume of the primary particles. While this approach may well represent coalesced droplets, it is mainly a simplified, convenient approach to represent any potential resultant agglomerate. The resultant equivalent particle actually represents the smallest possible size for the agglomerate and also the smallest possible surface area while still maintaining the conservation of mass of the colliding particles. Therefore, this approach will not be able to provide an indication of the resultant agglomerate structure.

We have developed a new approach to compute such an equivalent spherical particle yet be able to provide an indication on the looseness or compactness of the agglomerate (Jubaer *et al.*, 2019a). Building upon the capability to predict the degree of penetration of the colliding particle (Figure 4.6.1), the agglomerate surface area ratio can be computed as an indication of the degree of agglomeration between the particles A_g/A_{max}.

$$A_g = 4\pi R_1^{\,2} + 4\pi R_2^{\,2} - 2\pi R_1 h_1 - 2\pi R_2 h_2 \tag{4.6.1}$$

$$A_{max} = 2\pi R_1^2 + 2\pi R_2^2 \tag{4.6.2}$$

An equivalent particle can then be computed based on the agglomerate surface area as opposed to the conservation of the volume of the primary particles. On this basis, the A_g/A_{max} parameter can then be stored in memory and is updated upon each "sticking" agglomeration collision. One possible numerical scheme is given below. The final surface area ratio may range approximately from 0.8 to approaching 1. A value closer to 1 will indicate a looser agglomerate while a value closer to 0.8 will then indicate a more compact agglomerate formed. This routine is given in Figure 4.6.2. More details on the development of this concept are given in the references cited here

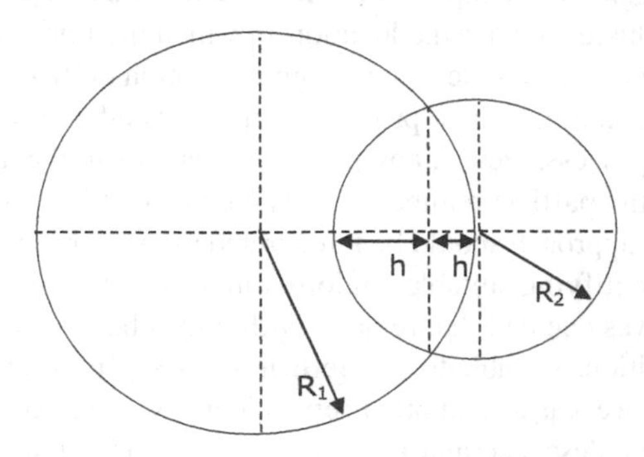

FIGURE 4.6.1 Visualization of the collision of two particles. (From Jubaer *et al.*, New perspectives on capturing particle agglomerates in CFD modelling of spray dryers. *Drying Technol.* 38 (5–6), 685–694, 2019a. With permission.)

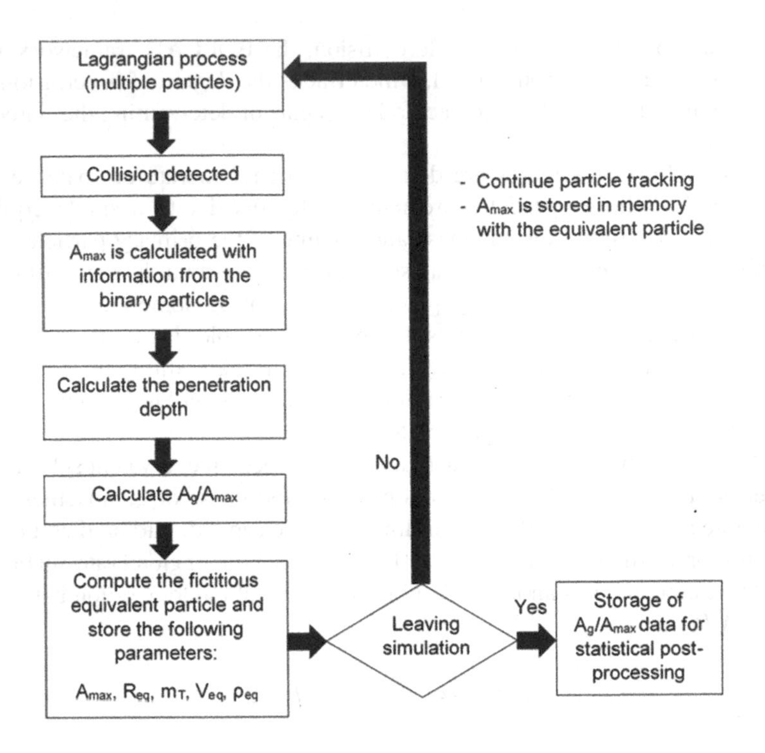

FIGURE 4.6.2 Numerical scheme for the implementation of the agglomerate surface area predictive approach in the Eulerian–Lagrangian CFD framework. (From Jubaer *et al.*, New perspectives on capturing particle agglomerates in CFD modelling of spray dryers. *Drying Technol.* 38 (5–6), 685–694, 2019a. With permission.)

and work is currently underway to evaluate this new approach to the computation of the equivalent agglomerate particle.

4.6.4 Agglomeration Mapping Approach

An alternative to agglomeration prediction may be to look at regions within the drying chamber to evaluate whether the location in which the fines and semi-dried particles are situated will lead to the desired agglomeration collision characteristics or not. This approach is mainly a post-processing method and does not explicitly model the agglomeration process. Hence, any potential changes in the agglomerate trajectory of changes in the particle number density in the chamber will not be captured. In other words, this approach should be interpreted as an approximation. Nestle used this approach in identifying suitable agglomeration regions based on the stickiness of the particles. It was deemed that regions within the chamber with sticky particles would lead to conditions suitable for the return of fines (Gianfrancesco *et al.*, 2009). In a similar vein, there was also another report identifying regions within the drying chamber prone to coalescence and agglomeration (Malafronte *et al.*, 2015). Their results, however, did not show any region suitable for coalescence, as that report was mainly focused on the development of their drying model. These reports, while able to identify regions that may be suitable for agglomeration, do not account for the effect of powder concentration on the agglomeration process.

We have developed an analysis method to determine suitable agglomeration regions within the spray drying chamber accounting for the rigidity of the powder as well as the particle number density (Jubaer *et al.*, 2019b). This analysis method was further verified and validated with experimental data from a laboratory-scale counter-current spray dryer, which showed the importance of accounting for the particle number density in such an analysis. More details can be found in the reference cited.

In these analyzes, as well as the explicit modeling of the agglomeration process, the use of the Ohnesorge number to distinguish between the droplet and the particulate state requires the viscosity property. In the authors' opinion, there is still a challenge to develop ways to measure or have a "continuous" description of the droplet viscosity from the liquid state to the near solid state. Most reports in the literature utilize the WLF concept and extrapolate the viscosity at the liquid state from the solid state by taking the glass transition temperature as the reference. We found that this approach may have a tendency to overpredict the viscosity of the atomized droplets leading to an unrealistic "no coalescence" prediction (Jubaer *et al.*, 2019b). An alternative approach would be to extrapolate the droplet viscosity, in reverse, from the experimentally measured viscosity at the concentrated liquid state to the solid-like state. Fortuitously, this approach was found to be more suitable in our analysis of the agglomeration of skim milk powder. Nevertheless, it will be of interest for future workers to examine and address this uncertainty in the prediction of droplet viscosity.

5 Monodisperse Droplet Spray Drying

5.1 WHY IS THIS CONCEPT IMPORTANT?

Based on the extensive experiences of the senior author of this book, Xiao Dong Chen, as an engineer working in Fonterra to produce milk powders and later as an academic working on industry-funded projects on spray drying in the period 1991–2003, he realized that it was difficult to draw definitive conclusions from the research results that were obtained using the conventional laboratory- or pilot-scale spray dryers at university or in companies, especially when the intention behind these studies was for scaling-up purposes and to optimize industrial operations. The small-scale dryers commercially available for laboratories can usually only produce very small particles, which come in all sizes and are of different morphologies, even from the same experimental run. It is usually not practical to carry out research on industrial-scale dryers, which are of the capacity of several to tens of tonnes of powder product produced per hour. In fact, the complexities encountered in the small laboratory dryers and the large industrial ones are similar, so the intention was to simplify the real situation to apply in a small laboratory. It is usually very difficult to discern the effects of operating parameters on morphological development among the particles made. To understand fundamentally what is going on in spray drying, a better-controlled scenario needed to emerge. In the period 2002–2003 at the University of Auckland, he proposed the making of a monodisperse droplet spray dryer (Patel, 2004). This was furnished through a number of PhD studies (Patel, 2008; Wu, 2010; Rogers, 2011).

5.2 DEVELOPMENT OF THE MONODISPERSE DROPLET SPRAY DRYER

5.2.1 Experimental Development over the Years

Although many modifications for better performance of the atomization system have been made in later years, the main components of the design have remained. Wu (2010) conducted the most significant study on the atomization behavior of the monodisperse droplet generators, proving there is a wide range of frequencies for the piezoelectric material to operate to generate uniform droplets. There are two modes of droplet generation through the piezoelectric mechanism shown in Figure 5.2.1a. A commercially available unit for continuous droplet generation mode is shown in Figure 5.2.1b. Figure 5.2.2 shows a commercially available monodisperse droplet spray dryer.

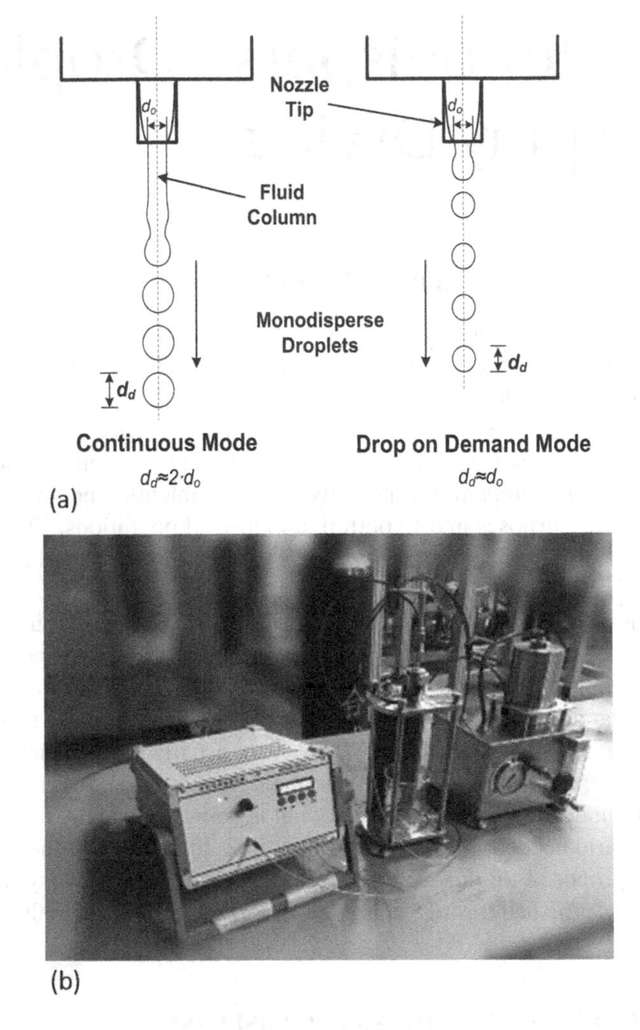

FIGURE 5.2.1 The principles of monodisperse droplet generation and the monodisperse droplet generation and demonstration unit available commercially. (Courtesy of Nantong Dong Concept New Material Ltd, China.)

The airflow pattern is simple and mostly straight downwards when the particles are relatively large: 100 microns and above. This helps make the particle trajectories simple to interpret quantitatively. The particles are easily collected under the dryer as the 'rainfall' of the particles can be easily captured due to the simple flow path. The uniformity of the particles made is excellent (refer to Figures 5.2.3 and 5.2.4) and the impact of operation parameters and feed characteristics can be more effectively studied.

The key to the success of the technology of the monodisperse droplet spray dryer is the ability to prevent or minimize the blockages occurring within the nozzle itself. There are some designs for the atomizer that have long thin capillary sections that make the pressure drop rather large, and a fibre or a particulate "contamination"

FIGURE 5.2.2 Monodisperse droplet spray dryer available commercially. (Courtesy of Nantong Dong Concept New Material Ltd, China.)

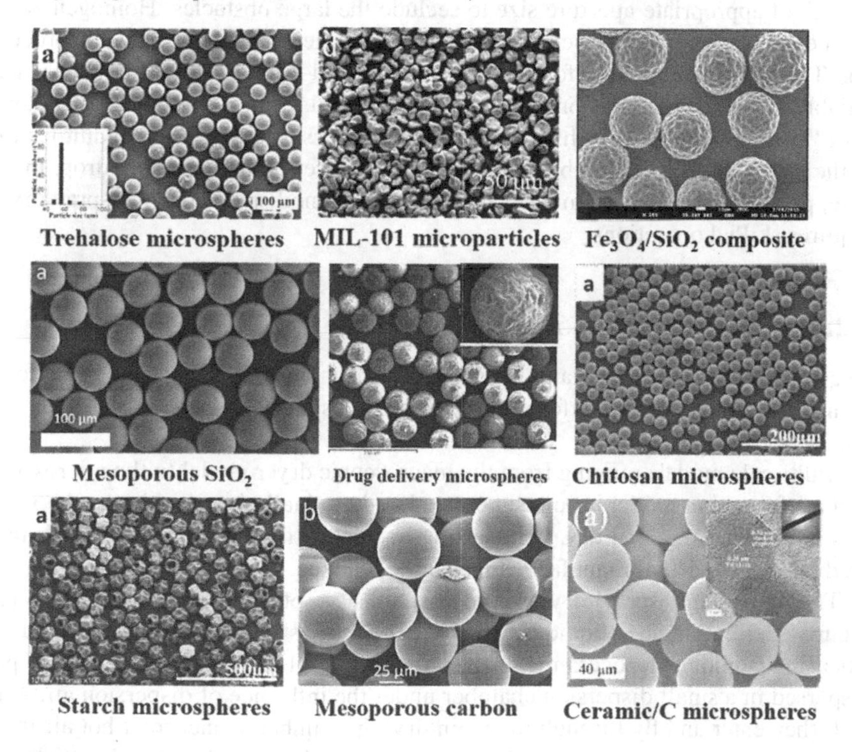

FIGURE 5.2.3 Various micro-particles produced using the device in Figure 5.2.2. (Courtesy of Nantong Dong Concept New Material Ltd, China.)

FIGURE 5.2.4 Uniform-sized whole milk particles produced the device in Figure 5.2.2. (Courtesy of Nantong Dong Concept New Material Ltd, China.).

within the fluid conduit leading to the nozzle tip can cause the whole process to fail. Sometimes, an inline filter needs to be implemented, such as that trialled by Rogers (2011). In general, it is recommended that the feed fluid should be filtered with a "sieve" of appropriate aperture size to seclude the large obstacles. Homogenization has been also commonly used to ensure the homogeneity of the fluid for processing. The nozzles fabricated for use by the Auckland–Monash track, now produced reliably at Nantong Dong Concept New Material Ltd, China, were those that have a very "shallow" nozzle tip which inserts a small pressure drop. The conduit leading to the nozzle tip is a "fat" tube that does not insert very much pressure drop. In any case, producing nice uniform particles using the monodisperse droplet spray dryers requires skilled operations.

5.2.2 Recent Computational Developments

Over the past decade, Xiao and his group at Soochow University have been developing a computational platform for the monodisperse droplet spray dryer (see Figure 5.2.2.1).

Multiscale models ranging from the macroscopic dryer model to the microscopic surface formation model have been created and gradually improved over the years. It is anticipated that one day, by resorting to this platform, "multiscale" "all-time" prediction, i.e., virtual manufacturing, can be realized.

The multistage spray drying process consists of the following stages (see Figure 5.2.2.2). The feed liquid is first atomized into a stream of uniform-sized droplets using a unique piezoceramic nozzle (Wu *et al.*, 2011). The droplets are then predispersed in a small dispersion chamber under the influence of dispersion air. After that, they enter and fly through the main drying chamber. Concurrent hot air in the dryer exchanges mass, energy, and momentum with droplets. The dried particles are finally collected at the outlet of the dryer.

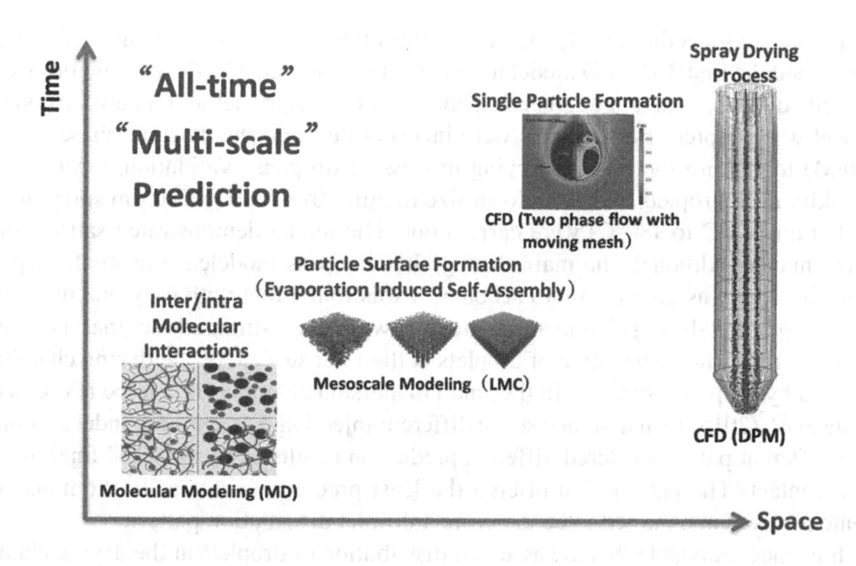

FIGURE 5.2.2.1 Multiscale modeling framework for the spray dryer. (Part of the diagram adapted from George, O.A., Xiao, J., Rodrigo, C.S., Mercade-Prieto, R., Sempere, J., Chen, X.D., Detailed numerical analysis of evaporation of a micrometer water droplet suspended on a glass filament. *Chem. Eng. Sci.* 165, 33–47, 2017.)

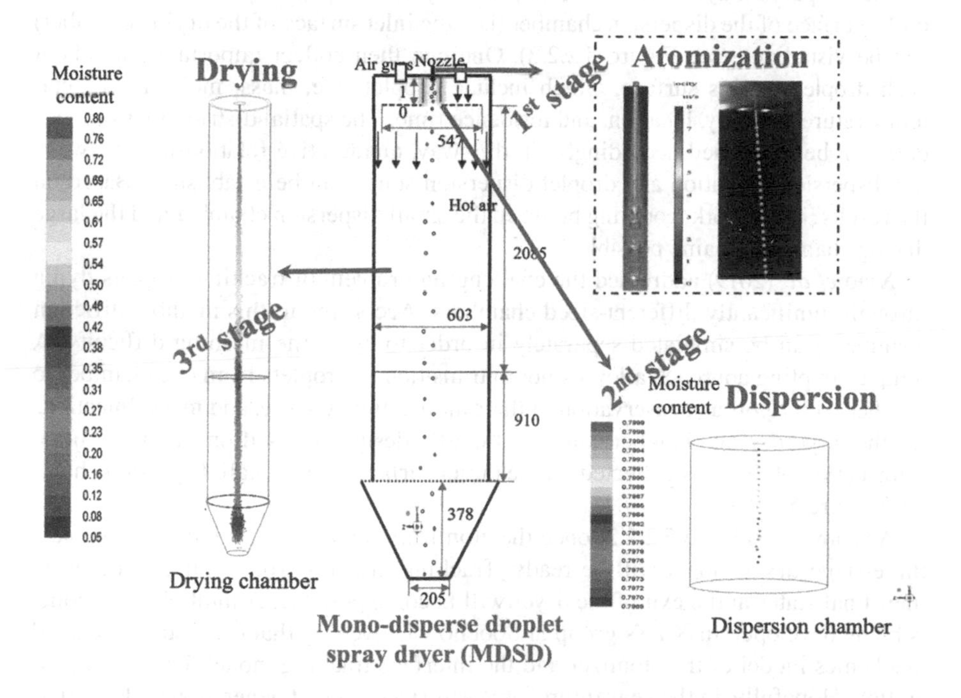

FIGURE 5.2.2.2 Macroscale CFD models for the multistage drying process. (Adapted from Xiao, J., Yang, S.J., George, O.A., Putranto, A., Wu, W.D., Chen, X.D., Numerical simulation of monodisperse droplet spray dryer: Coupling distinctively different sized chambers. *Chem. Eng. Sci.* 200, 12–26, 2019.)

In order to track droplets flying through this unique spray dryer, Yang *et al.* (2015) developed the first 3-D CFD model for the drying chamber. The drying kinetics were described by the reaction engineering approach (REA) model and a new shrinkage model was adopted. Both models were incorporated into the discrete phase model (DPM) to capture the realistic drying process of droplets. Validation experiments for skim milk droplets with a uniform size ranging from 180 to 220 μm spray dried by hot air (90°C to 180°C) were carried out. The model demonstrated satisfactory performance. Although the main drying chamber was modeled, the small dispersion chamber was not taken into account. It was found that with only one injection at the top of the drying chamber, the model always overestimated the final moisture content. Thus, the distribution of droplets at the inlet surface of the drying chamber (caused by droplet dispersion in the small dispersion chamber) had to be addressed. Yang *et al.* (2015) then assumed seven different injection patterns and, indeed, found that different patterns offered different prediction results as to particles' final moisture content. The pattern that offered the least prediction error as to the moisture content was then assumed to be the correct droplet distribution pattern.

It is understandable that the assumed distribution of droplets at the drying chamber inlet may not correspond to the realistic distribution. Since the status of droplet distribution can hardly be observed in experiments, Xiao *et al.* (2018a) carried out numerical simulations for this dispersion process. It allowed them to capture the droplet dispersion dynamics. More importantly, the droplet dispersion states on the outlet surface of the dispersion chamber (i.e., the inlet surface of the drying chamber) can be visualized (see Figure 5.2.2.3). One can then collect important data about each droplet on this surface, which include droplet size, mass, moisture content, temperature, velocity, location, and residence time. The spatial distributions of these data can be quantified accordingly. In this way, quantitative relationships between the dispersion operation and droplet dispersion states can be established. Based on the results in this work, coupling between the small dispersion chamber and the large drying chamber became possible.

Xiao *et al.* (2019) addressed the challenging problem of tracking droplets flying through significantly different-sized chambers. According to this method, different chambers can be simulated separately in order to avoid the meshing difficulty. A unique coupling approach allows smooth transition of droplets from one chamber to another with rigorous conservation of their momentum, energy, and mass. Injections for the large drying chamber can be rationally designed based on the distribution information of droplets collected on the outlet surface of the small dispersion chamber (Figure 5.2.2.3).

As shown in Figure 5.2.2.2, once the atomization model is available, a complete three-stage dryer model will be ready. Tracking droplets from their generation to their final states at the exit of the dryer will become possible. A multiphysics model is being developed in Xiao's group at Soochow University that can couple the solid mechanics model of the atomizer and the interface tracking model for droplet generation. Hopefully, in the near future, uniform-sized droplet generation under different atomization conditions can be simulated.

The above-listed CFD simulations assumed spherically shaped particles in the DPM model. Differently shaped particles, however, are frequently encountered in

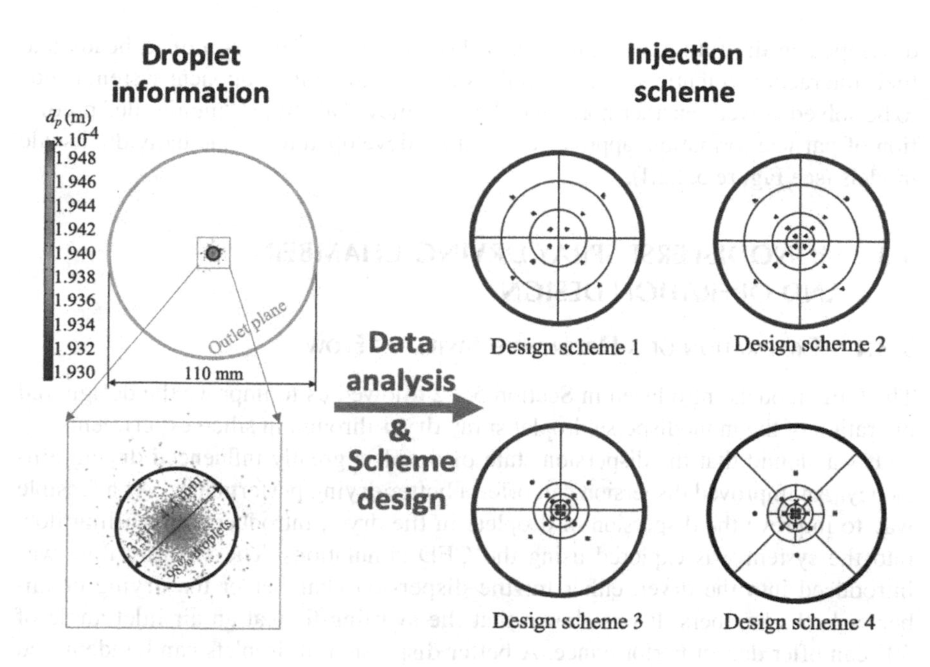

FIGURE 5.2.2.3 Illustration of the approach to couple the small dispersion chamber and the big drying chamber. The shaded dots represent droplets on the outlet surface of the distribution chamber. The shading indicates droplet diameter. This method is an effective way to model discrete phase particles flying through distinctively different-sized chambers. (Adapted from Xiao, J., Yang, S.J., George, O.A., Putranto, A., Wu, W.D., Chen, X.D., Numerical simulation of monodisperse droplet spray dryer: Coupling distinctively different sized chambers. *Chem. Eng. Sci.* 200, 12–26, 2019.)

a real powder production process. Prediction of droplet shape evolution during the drying process is a very challenging task that has not been accomplished so far. Complex transport phenomena, together with phase change, property change, and even chemical reactions occurring inside and outside the droplet, have to be taken into account. George *et al.* (2017) made a first-step attempt to simulate the evaporation of a 1 μm pure water droplet suspended on a glass filament. A two-phase flow model was developed, where the evolution of the droplet–air interface was tracked by the moving mesh method (see Figure 5.2.2.1). This model was validated using experimental data collected from our unique single droplet drying facility. To investigate particle formation, the model was extended recently to simulate the drying of a lactose droplet (George *et al.*, 2019). In the model, highly viscous fluid was used to represent the solid particle. How to simulate the formation of particles with complicated shapes and structures remains an unsolved problem.

In addition to the dryer scale and single droplet scale models, Xiao's group also pursued molecular-scale models that can offer the surface structure of spray dried particles. Shang *et al.* (2019) developed a coarse-grained lattice Monte Carlo model that can simulate evaporation-induced self-assembly of solute molecules. Although the model can take care of different-sized solute molecules, it is just the very earliest attempt. The model is currently for 2-D systems only. A 3-D model should be developed in the future. The problem of how to design coarse-grained beads and their interaction potentials for the simulation of a real multicomponent system needs to be solved as well. Furthermore, in order to achieve "all-time" "multiscale" prediction of particle formation, approaches must be developed to couple individual scale models (see Figure 5.2.2.1).

5.3 MONODISPERSE SPRAY DRYING CHAMBER AND OPERATION DESIGN

5.3.1 Exploration of a Dryer with Swirling Flow

The CFD models introduced in Section 5.2.2 allowed us to improve the design and operation of the monodisperse droplet spray dryer through in silico experiments.

It was found that the dispersion state of droplets greatly influenced drying efficiency. An improved dispersion can offer a better drying performance. As a feasible way to improve the dispersion of droplets in the dryer, introducing a swirling flow into the system was explored using the CFD simulations. The swirling flow was introduced into the dryer, either for the dispersion chamber or the drying chamber or both chambers. It was shown that the swirling flow at an air inlet angle of 30° can offer decent performance. A better dispersion of droplets can be identified as compared with the case without any swirling flow (see the comparison between Figure 5.3.1a and Figure 5.2.2.2), especially for the top section of the dryer with high turbulent kinetic energy (see Figure 5.3.1b). When introducing swirling flows into both chambers, the co-current scheme was slightly better than the counter-current scheme in terms of reaching a lower particle moisture content. The swirling flow could significantly improve drying efficiency. The moisture content of the particle

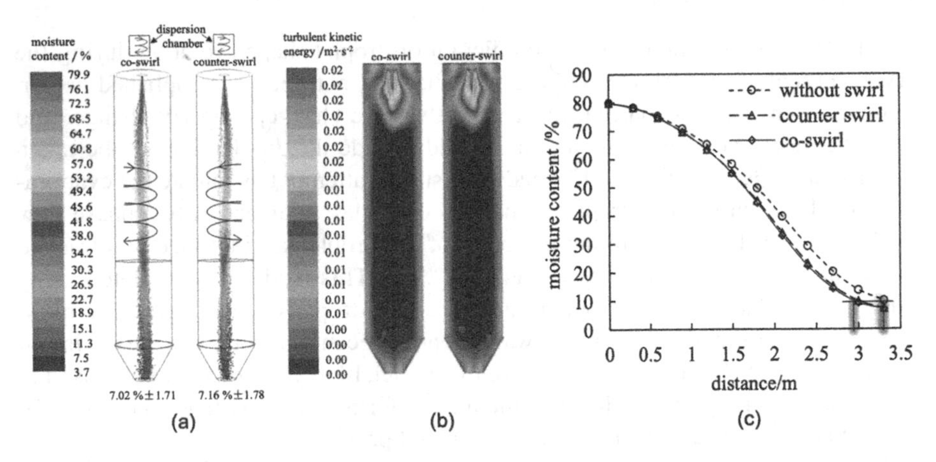

FIGURE 5.3.1 Simulation of a monodisperse droplet spray dryer with swirling flow: (a) distribution of droplets in the dryer; (b) turbulent kinetic energy; (c) evolution of particle moisture content. (Adapted from Yang, S.J., Wei, Y.C., Woo, M.W., Wu, W.D., Chen, X.D., Xiao, J., Numerical simulation of monodisperse droplet spray dryer under influence of swirling flow. *CIESC J.* 69(9), 3814–3824, 2018.)

product can be 30% lower when compared with the case without any swirling flow. A shorter dryer (of nearly 12% reduction in height) could achieve the same moisture content with the introduction of the swirling flow (see Figure 5.3.1c).

5.3.2 Exploration of a Dryer with Moving Atomizer

Another interesting effort concerning the improved design of dryers is to implement a moving atomizer, i.e., a moving nozzle. It is understandable that a better dispersion can be achieved through nozzle motion. The design space that includes the trajectory and speed of motion, however, is not small. In silico experiments can significantly facilitate and speed up the design process.

Using the CFD model developed in our group, Wei *et al.* (2019) systematically tested three different types of motions, i.e., linear, circular, and sinusoidal motions. Wider distributions of droplets were observed when compared with the case without nozzle motion (see Figure 5.3.2). It was found that for linear and sinusoidal motions, particles with relatively high moisture content were mainly concentrated on the central and terminal points of movement trajectories. The circular motion was found to be the best design, since it offered the highest drying efficiency and the narrowest distribution of moisture content among all motion types. It was confirmed that the nozzle movement promoted a better dispersion of droplets and weakened the interaction between them, which is the major mechanism for improved drying performance.

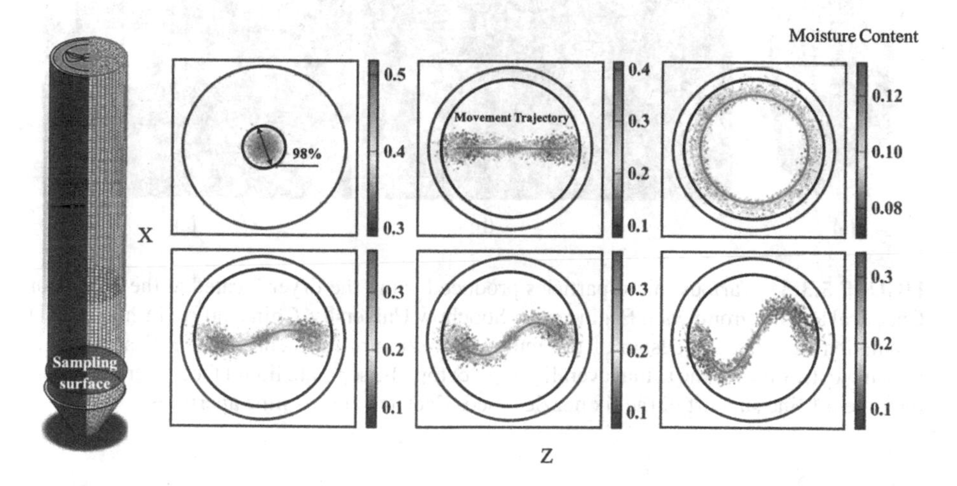

FIGURE 5.3.2 Simulation of a monodisperse droplet spray dryer with a moving atomizer. The properties of droplets passing through the sampling surface are recorded. Droplet distributions on the sampling surface under different nozzle motion types are plotted for comparison. The dots represent droplets and the brown lines display the movement trajectories of the nozzle. The shading of a dot indicates the moisture content. (Adapted from Wei, Y.C., Woo, M.W., Selomulya, C., Wu, W.D., Xiao, J., Chen, X.D., Numerical simulation of monodisperse droplet spray dryer under the influence of nozzle motion. *Powder Technol.* 355, 93–105, 2019.)

5.3.3 Scale-Up of Monodisperse Droplet Spray Dryer

One of the scaled-up versions of the monodisperse droplet spray dryers, producing some 200–1000 g of powder per hour, is shown in Figure 5.3.3. Essentially, on this dryer, there are 54 "nozzle tips" working simultaneously to deliver the liquid into the dryer chamber.

For silica-based particles, this dryer was very reliable and particle uniformity was excellent as well. However, blockages were also evident at some of the nozzle tips, and strategies to improve this have been thought of but need further R&D. Successful implementation of about 1 kg per hour capacity or greater will bring the technology right into the legitimate range for the pharmaceutical industry or other high-value particle product industries.

FIGURE 5.3.3 Various micro-particles produced using the dryer located at the School of Chemical and Environmental Engineering, Soochow University, China. (a) An in-house CFD simulation showing the design of the multiple nozzle arrangement and the dispersing air blowing outwards; (b) the actual overall setup; (c) top: the separate liquid feed reservoirs; bottom: the off-line support of the six nozzles used. Each nozzle has nine apertures.

6 Advanced Applications of Spray Drying

6.1 SUPERHEATED STEAM SPRAY DRYING

Superheated steam drying has been well established for many forms of solid-based drying. Application of superheated steam as the drying medium in wood, sludges, food, and agricultural products have shown that the medium improves product qualities such as porosity, and prevents and minimizes undesired quality changes such as off-flavors and oxidation reactions. There are, however, minimal scientific reports on the use of superheated steam as a medium for spray drying. Early reports in the 1960s and 1970s evaluated and focused on the heat and mass transfer characteristics of superheated steam drying of droplets. Beyond those fundamental works, there have been minimal reports extending these to laboratory-scale or pilot-scale dryers. The series of simulation works by Frydman and co-workers provide more insights into the heat and mass transfer aspect of superheated steam in a pilot-scale spray dryer (Frydman *et al.*, 1998). There was, however, no reported experimental work in that report. In this section, some recent experimental work in this area will be discussed with more emphasis on how superheated steam can be used in spray dryers with reference to its effect on powder functionality development. The subsequent laboratory-scale superheated steam spray drying work was undertaken using the counter-current superheated steam spray dryer illustrated (Figure 6.1.1). In addition, we have also undertaken experimental investigation. These rigs and the intricacies associated with the use of superheated steam will be discussed in detail later on (Figure 6.1.2).

6.1.1 IMPROVING THE WETTABILITY OF DAIRY POWDER

It is well known that whole milk particles are overrepresented on the droplet surfaces by fat, due to the surface active mechanism and the diffusion mechanism leading to the migration of fat to the droplet surface during spray drying. An interesting aspect of the initial experimental work using the single droplet drying technique was that superheated steam drying led to single droplet dried particles with significant improvement in wettability behavior (Lum *et al.*, 2017). In some of the experimental work, the contact angle of the particle was reduced by approximately half of that dried by hot air. Figure 6.1.3 illustrates this significant reduction in contact in a set of additional controlled experiments coating a plate surface with whole milk dried under superheated steam conditions versus hot air. The purpose of the flat plate control was to illustrate this phenomenon in the absence of any potential differences in the curvature of the dehydrated particle via the single droplet rig. Initial experimental work with the single droplet technique revealed that superheated steam resulted

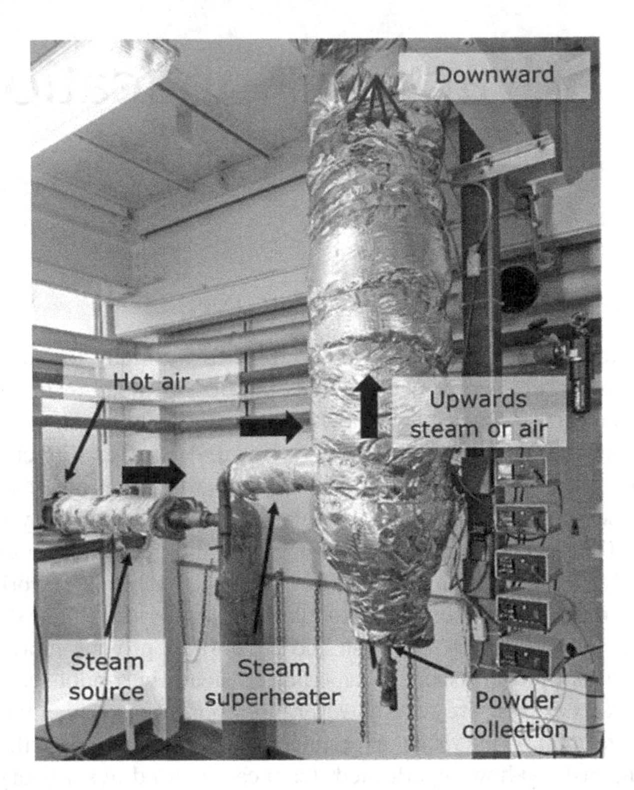

FIGURE 6.1.1 Laboratory-scale counter-current superheated steam spray dryer. (From Woo, M.W., Advances in production of food powders by spray drying, in *Advanced Drying Technologies for Food*, edited by Mujumdar, A.S., Xiao, H.W. Taylor & Francis, Boca Raton 2019. With permission.)

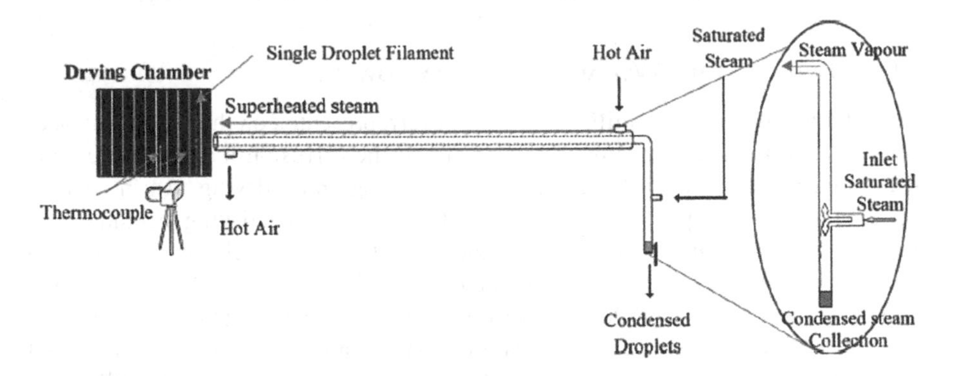

FIGURE 6.1.2 Superheated steam single droplet drying rig. (From Lum, A., Mansouri, S., Hapgood, K., Woo, M.W., Single droplet drying of milk in air and superheated steam: Particle formation and wettability. *Drying Technol.* 36, 1802–1813, 2017. With permission.)

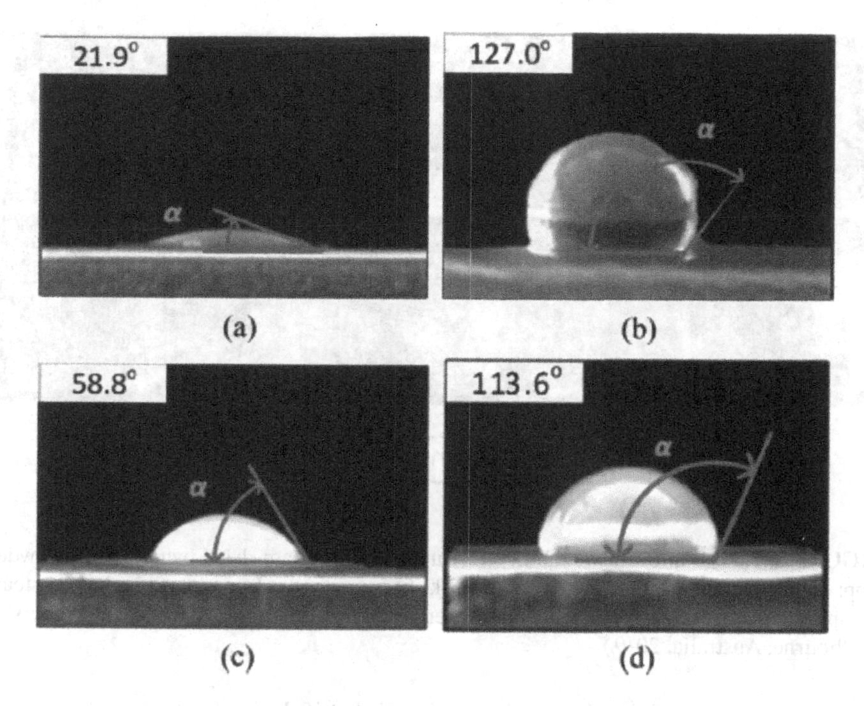

FIGURE 6.1.3 Improved wettability of whole milk film dried with superheated steam. (From Lum, A., Mansouri, S., Hapgood, K., Woo, M.W., Single droplet drying of milk in air and superheated steam: Particle formation and wettability. *Drying Technol.* 36, 1802–1813, 2017. With permission.)

in the slower formation of the crust typically associated with the formation of hollow whole milk particulates under convective dehydration.

Although it may appear that superheated steam drying led to a different surface composition, which could have led to a different contact angle, samples from the single droplet drying technique did not facilitate particulate surface characterization. The study was then extended to laboratory-scale spray drying using the counter-current superheated steam spray dryer. A similar observation was found in which the superheated steam-dried whole milk powder exhibited significantly higher wettability (Lum, 2019). Figure 6.1.4 shows the wettability evaluation undertaken compared with hot air-dried whole milk powder. Surprisingly, XPS analysis of the powder revealed no significant difference in the surface fat of the powder. From further SEM analysis, it was deduced that this better wettability may be due to the slightly more porous surface morphology in the superheated steam-dried whole milk powder.

6.1.2 A Medium for In-Situ Crystallization Control

Experiments have shown that superheated steam is a useful medium to control the in-situ crystallization process of fast to crystallize materials in spray drying (Lum *et al.*, 2018, 2019). A more detailed definition of the in-situ crystallization process is given

FIGURE 6.1.4 Improved wettability of superheated steam-dried whole milk powder. Top: air-dried milk; bottom: steam-dried milk. (Adapted from Lum, A., Superheated steam in spray drying for particle functionality engineering. PhD thesis, Monash University, Melbourne, Australia, 2019.)

in the next section of the chapter. It was found that if drying at the same inlet temperature, superheated steam will induce higher rates of nucleation when compared with drying using hot air. The higher nucleation rate further translates to finer crystal grains forming individual granular particles and vice versa. This is directly linked to and can be explained by the different drying history of superheated steam drying.

A droplet drying in hot air inevitably experiences and goes through the wet bulb region to a certain extent, depending on its propensity to form a crust during the drying process. In any case, the droplet will start to experience significant solidification (or formation of a surface crust) at about the wet bulb temperature (Figure 6.1.5). Under superheated steam drying conditions, however, the droplet will rapidly reach the saturation or boiling temperature corresponding to the pressure of the superheated steam. Within the range of air humidity used in typical spray drying operations, the saturation temperature will be significantly higher than the wet bulb temperature. In effect, the droplet will then experience solidification at a relatively higher temperature. As crystallization is, in essence, occurring during the solidification of the droplet, droplets under a superheated steam environment will undergo in-situ crystallization at a higher temperature, which will directly influence and increase the nucleation rate of the crystallization process.

Achieving such relatively high "solidification" or crystallization temperature with hot air will be relatively difficult, as this involves either using very humid air, which will potentially impede drying, or requires the use of an excessively high inlet temperature, which may be detrimental to the quality of the product. Such an excessively high inlet temperature may also lead to very rapid dehydration or quenching, limiting the potential for the development of the crystals in the dehydrating droplet.

In addition to producing crystal particles of different sizes, high nucleation may also change the morphology of the crystals produced. Figure 6.1.6 illustrates the

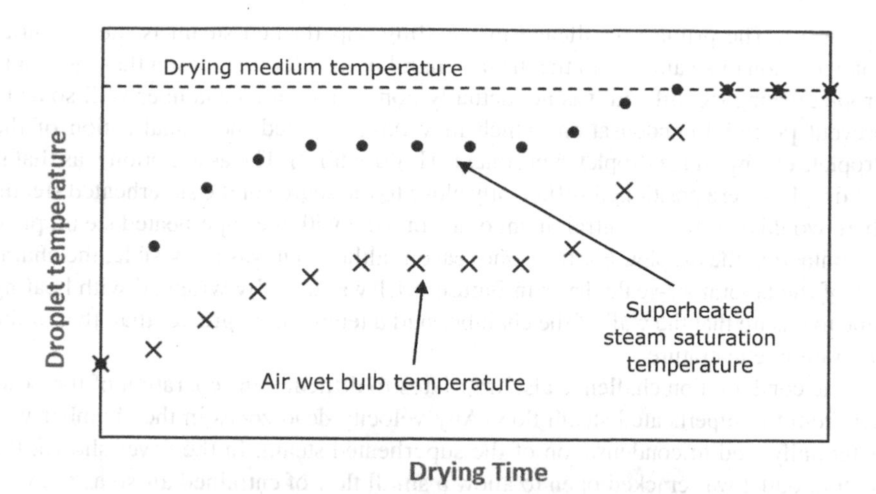

FIGURE 6.1.5 Droplet drying history under superheated steam and air environment.

FIGURE 6.1.6 Bipyramidal salt particles from superheated steam spray drying; particle size is approximately 200 microns. (Adapted from Lum, A., Cardamone, N., Beliavski, R., Mansouri, S., Hapgood, K., Woo, M.W. The role of steam as a medium for droplet crystallization. *Ind. Eng. Chem. Res.*, 58, 8517–8524, 2019.)

unique bipyramidal salt particles produced from intensified secondary nucleation of the salt droplets under superheated steam drying conditions, in contrast to the traditional cubic crystalline structure in typical salt.

6.1.3 Important Technical Considerations When Using Superheated Steam

This section touches on a few important notes for future workers in this area, highlighting the challenges when using superheated steam for spray drying in contrast to hot air. The primary challenge in handling superheated steam is the potential condensation of steam within the drying chamber. It is for this reason that the single droplet drying rig utilized was not actually contained with a chamber wall so as to prevent potential condensation, which may have affected the visualization of the droplets during single droplet experiments (Figure 6.1.2). The assumption was that if the droplets were positioned sufficiently close to the source of the superheated steam, there would be minimal entrainment of air mixing with the superheated steam prior to contacting the droplet. For the same reason, although it was not visible, the chamber of the laboratory-scale dryer in Figure 6.1.1 was actually wrapped with heating tape to ensure that the wall of the chamber had a temperature greater than that of the saturation temperature.

The condensation challenge also translates to difficulty in separation of the powder from the superheated steam flow. Any velocity dead zones in the chamber will potentially lead to condensation of the superheated steam. In the dryer shown, the bottom outlet was cracked open to allow a small flow of entrained air so as to avoid this condensation problem; this allowed the powder to be collected by gravity against this slight upward ingress of air. Similar challenges are also encountered if conventional cyclone powder removal is used. The team at King Mongkut's University of Technology Thonburi (KMUTT) has done extensive work developing a powder removal system at the bottom of their cyclone for a modified laboratory-scale spray dryer unit (Feungfoo *et al.*, 2018).

For the single droplet rig and the laboratory-scale dryer used, another important aspect of this form of superheating used was the need for a T-shaped pipe (or other equivalent system) to remove saturated steam droplets preceding superheating. This ensures that the steam can be effectively superheated. While this aspect will be accounted for in commercial superheater units, such a strategy will be important for future workers (particularly for laboratory applications) interested in the development of their own superheater. Acknowledgement is given to Professor Sakamon Devahastin (KMUTT) for this critical advice on the need for steam droplet separation.

6.2 CONTROLLING IN-SITU CRYSTALLIZATION IN SPRAY DRYERS

6.2.1 General Overview of In-Situ Crystallization

This section of the book focuses on the control of crystallization of the dissolved solids in the droplets during the dehydration process. Most reports on the control of crystallization in spray dried powder focuses on analysis during post-drying storage or under specific storage-tempering treatment under different humidities or temperatures. These are not the focus of this section of the book. In addition, there are also numerous reports in the literature focusing on how different product formulation affects the crystallization behavior of droplets during the spray drying process. In contrast, the purpose of this section of the book is to provide a deeper discussion on how the spray drying process can be manipulated to control crystallization in a process mechanistic approach. It is hoped that the technique described can be applied across different product formulations. Table 6.2.1 summarizes some of these process mechanistic-based approaches.

TABLE 6.2.1
Summary of Mechanistic Papers on In-Situ Spray Crystallization Control

References	Materials	Type of spray dryer	Main control strategy	Purpose of control
Littringer *et al.* (2012)	Mannitol	Pilot-scale	Air inlet temperature and the liquid feed rate	To control particle surface characteristics
Mass *et al.* (2012)	Mannitol	Pilot-scale	Air outlet temperature	To control particle surface characteristics
Littringer *et al.* 2013	Mannitol	Mobile Minor Tall form pilot-scale	Adjustment of the droplet size by using different atomizers Adjustment to the outlet temperature	To evaluate the different morphology, sizes, and crystallinity of the particles
Islam *et al.* (2010)	Lactose	Buchi lab-scale	Using high humidity within the spray dryer	To induce in-situ crystallization of spray dried lactose
Islam and Langrish (2010)	Lactose	Buchi lab-scale	Controlling the inlet temperature to very high levels	To induce in-situ crystallization of spray dried lactose
Chidavaenzi *et al.* (1998)	Lactose	Buchi lab-scale	Controlling the inlet liquid feed temperature	To control the polymorphic form of the amorphous particles during drying
Buckton *et al.* (2002)	Lactose	Buchi lab-scale	Controlling the inlet liquid feed temperature	To control the polymorphic form of the amorphous particles during drying
Littringer *et al.* (2013)	Mannitol	Tall form pilot-scale dryer (3.7 m high × 2.7 m diameter)	Manipulating the product outlet temperature	To understand how the outlet temperature affects particle surface morphology due to changes in crystallization behavior
Das and Langrish (2012) Das *et al.* (2010)	Lactose	Pilot-scale spray dryer	Controlling the inlet temperature, hot air flow rate, atomization rate (in the latter paper, incorporation of additional hot air near the atomizer region)	To understand how longer residence time in a larger-scale spray dryer affects the control strategy for in-situ crystallization
Lin *et al.* (2015)	Glycine	Mono-disp. spray tower	Manipulating the inlet air temperature	To assess how different evaporation rates lead to hollow or semi-hollow crystalline structure

(*Continued*)

TABLE 6.2.1 (CONTINUED)
Summary of Mechanistic Papers on In-Situ Spray Crystallization Control

References	Materials	Type of spray dryer	Main control strategy	Purpose of control
Sander *et al.* (2011)	Glycine	Buchi lab- scale	Manipulating air temperature, feed concentration, feed rate, and airflow rate	To understand the effect of various operating parameters on glycine polymorphism
Ebhrahimi and Langrish (2015)	Lactose	Buchi lab-scale	Controlling humidities by providing humid air before the main spray dryer	To study the effect of different humidities on process yields and degree of crystallinity
Shakiba *et al.* (2019)	Lactose Mannitol	Counter-current spray dryer	Controlling the feed temperature and counter-current hot air temperature	To understand how the counter-current drying history affects the in-situ crystallization process
Mansouri *et al.* (2015)	Sodium chloride Sucrose Lactose	Narrow tube spray dryer	Controlling the residence time and the air temperature throughout the entire droplet drying history	To assess how continuous heating of the air (the drying history of the droplet) affects the in-situ crystallization process

Source: Adapted from Woo, M.W., Lee, M.G., Shakiba, S., Mansouri, S., Controlling in situ crystallization of pharmaceutical particles within the spray dryer. *Expert Opin. Drug Deliv.* 14, 1315–1324, 2017.

To better interpret the subsequent discussion, it is important to clearly define what in-situ crystallization control means, depending on the material of interest. For this purpose, broadly, spray dried materials can be classified into easy to crystallize and difficult to crystallize materials. Most spray dried food powders are typically difficult to crystallize during spray drying. This is mainly because of the rapid solidification that quenches the solute into the amorphous form, preventing effective crystallization. Some examples of difficult to crystallize materials are lactose, sucrose, starches, coffee powder, etc. For such materials, the control of crystallization mainly pertains to generating partial crystallinity; the focus is on controlling the percentage of solids within the powder in the crystalline form (or the relative percentage between batches of powder). While there is continuous scientific interest in controlling or increasing the percentage of crystallization within such typically amorphous powder, partial crystallinity may not be beneficial for commercial production of these powders. This is because partial crystallinity may actually provide the seed for undesired phase changes during the long-term storage of the typically amorphous powder. Generating partial crystallinity, however, has been widely adopted in the spray drying of whey powder to control the stickiness of the powder. In the spray drying of whey powder, the feed solution is typically allowed to undergo cooling crystallization so that part of the lactose is crystallized prior to spray drying.

In contrast, salt materials and some pharmaceutical excipients such as mannitol are typically produced in the crystalline form. In-situ crystallization then specifically refers to the control of the crystalline structure within the particles and not the control of the percentage of crystallization within the particle. In particular, there is continuing interest in controlling the size of the ultrafine crystal precipitates, which agglomerate to form the whole particle. Depending on the operation of the spray dryer, the polymorph of the crystals can also be manipulated.

6.2.2 Strategies for Materials That Are Slow to Crystallize

The strategies described in this section pertain mainly to the spray drying of non-seeded (fine crystals) feed solutions to the spray dryer. For materials that are hard to crystallize, there are currently two main schools of thought for in-situ crystallization control. There are many reports – in fact, most of the lactose work in Table 6.2.1 alludes to the formation of partial crystallization within the solid phase transition of the material. Within this theoretical framework, the droplets would have been rapidly dehydrated and quenched into the amorphous form. The amorphous particles then undergo phase transition above the glass transition state into the partial crystalline form. It is noteworthy that this is well established, particularly in the long-term storage of amorphous powder. The reports on lactose in Table 6.2.1 suggest that the same phenomenon can occur within the short time scale in the spray drying chamber.

For solid phase transition to occur, it is pertinent that the powder be maintained in sufficient high mobility condition beyond the glass transition state throughout its trajectory within the spray chamber. It is well known that the outlet conditions from the spray chamber are representative of the conditions within the chamber (in contrast to the inlet air conditions). Assuming equilibrium between the powder and the outlet conditions, one proposed framework is to maintain the outlet temperature to be higher by 20–30°C above the glass transition temperature of the powder to ensure high molecular mobility so as to induce solid phase transition in the powder. This can be achieved by using a higher inlet temperature, which will lead to a higher outlet temperature. The degree of elevation of the inlet temperature has to be balanced by the inevitable reduction in the moisture content of the powder, which will, in contrast, increase the glass transition temperature of the powder. It is noteworthy that under this framework, the positive difference beyond the glass transition temperature denotes the degree of solid phase transition (crystallization). Alternatively, there are suggestions to reduce the glass transition temperature by using more humid inlet gases, which effectively increases the moisture content of the powder. Similarly, this approach will have to be balanced against insufficient drying of the powder, which will inevitably lead to a stickiness problem within the chamber, further affecting the stability of the spray dried powder.

Regardless of which approach is adopted, the solid phase transition framework indirectly implies that crystallization occurs towards the later stages of the particle trajectory, during the period in which the particle is already formed. This framework does not specifically distinguish between nucleation and growth but focuses on maximizing the residence time and the "intensity" in which the particles remain above the glass transition state within the drying chamber. The outlet condition glass transition-based control framework was developed mainly in small-scale spray dryers. There is some evidence in the literature to suggest that such a control strategy may need to be augmented for large-scale dryers (when scaling up) (Das *et al.*, 2010; Das and Langrish, 2012).

In contrast to controlling the outlet condition of the spray drying chamber, there is a series of works focusing only on intensifying the critical period of crystallization during the trajectory of the droplet. This concept was firstly proposed by Woo *et al.* (2012), analysing crystallizing lactose droplets with single droplet experiments. Even in the absence of a defined crystallization mechanism, it was found that the period approximately corresponding to that just before the increase in droplet temperature from the wet bulb period to a period just after the droplet reaches the ambient drying air temperature is critical in the control of in-situ crystallization. In contrast, relative to this critical period, the potential solid phase transition towards the latter stages of the particle drying history may be insignificant. This theoretical framework was further evaluated using a counter-current spray dryer (Esfahani *et al.*, 2018). In those experiments, the critical crystallization period was extended by introducing a secondary water spray at locations corresponding to the critical crystallization region. The aim was to delay the drying rate at the critical zone to extend the time duration available for in-situ crystallization. This strategy led to the production of very highly crystalline lactose powder.

The report also revealed a strategic aspect of counter-current spray drying for in-situ crystallization control using the high humidity approach. In those experiments, as the primary was atomizing downwards while the counter-current air was flowing upwards, any introduction of high humidity only affected the upstream and initial trajectory of the powder and did not directly affect the latter stages of drying. In contrast to the use of a co-current dryer, the introduction of high humidity will inevitably affect the drying history during the latter trajectory of the powder, which will make the powder more prone to insufficient drying. Therefore, a more local in-situ crystallization humidity control can be achieved with the counter-current spray tower technique.

6.2.3 Strategies for Materials That Are Fast to Crystallize

For materials that are fast to crystallize, there were similar suggestions in the literature to use the drying chamber outlet conditions as a means to delineate the drying behavior to further control the crystallization of the material. Most of the reported work in this area, as shown in Table 6.2.1, was undertaken with mannitol. As discussed earlier, the primary purpose of crystallization control was to manipulate the morphology of the crystalline structure, specifically the size of the individual crystal grains forming the overall powder particulate, which is governed by the combination of nucleation and growth rates. There are contrasting results at different dryer scales as to how the outlet temperature controls the size of the crystal grains. We opine that this approach by which to control the crystalline powder morphology may not be suitable.

Vehring (2008), in a series of works, introduced the theoretical framework that directly controlled the time duration within the trajectory of the droplet that is available for the crystallization to occur. Using the Peclet number as the basis, this available period is denoted by the time in which the solute concentration at the surface of the droplet exceeds that of saturation; a basic prerequisite for nucleation and growth. This approach was shown to be able to delineate and describe the in-situ crystallization behavior of amino acid particles. One gap in that original framework was that while the framework captured the effect of the drying conditions on the surface solute enrichment of the droplet, it did not differentiate or capture the crystallization behavior of the solute. Solutes are mainly differentiated by their diffusivity behavior in the solvent, which is not directly related to the crystallization behavior. In other words, the framework does not distinguish between the nucleation and growth of the material throughout the in-situ crystallization process.

Building upon the framework by Vehring, Shakiba *et al.* (2019) have developed the use of a Damkohler number to differentiate the time duration available for nucleation and growth of the dissolved solutes. The theoretical control framework visualizes the crystallization process as a "reaction" process and evaluates its time scale relative to the overall evaporation time scale. This approach effectively incorporates the crystallization behavior of the solute in the predictive framework. The predictive framework was able to describe the in-situ crystallization behavior of mannitol powder in the same counter-current spray tower used in the lactose crystallization study described earlier.

The work by Shakiba on the control of in-situ crystallization of fast to crystallize materials also revealed an interesting aspect of counter-current spray drying. The spray drying of mannitol, under certain drying conditions, led to "Janus-like" particles in terms of its crystalline morphology (Figure 6.2.1). This is clearly an artefact of significantly different local drying behavior on different locations on the droplet surface. Such morphologies have never been reported before in any work utilizing co-current spray dryers at different scales. We speculate that such differences are

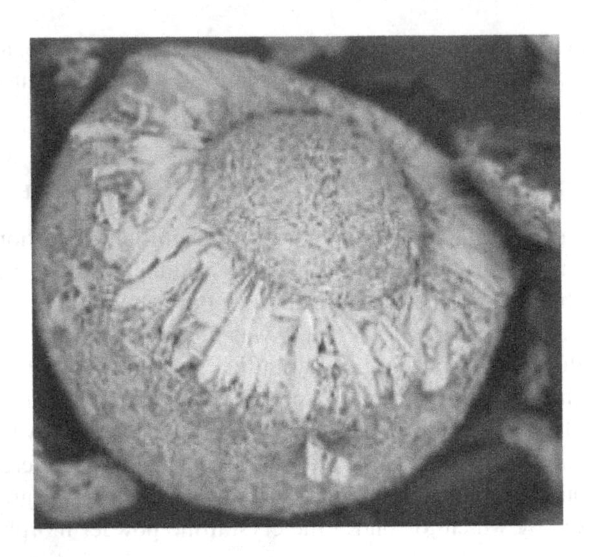

FIGURE 6.2.1 Janus-like mannitol particles. (From Shakiba, S., Mansouri, S., Selomulya, C., Woo, M.W., Time scale based analysis of in-situ crystal formation in droplet undergoing rapid dehydration. *Int. J. Pharm.* 50, 47–56, 2019. With permission.)

indeed inherent in spray dried droplets but this may be more significant, or not, depending on the rate of drying within the chamber. Figure 6.2.2 illustrates the fundamental difference in the drying history between co-current and counter-current spray dryers. It can be seen that co-current spray dryers have a very rapid initial drying and solidification rate, while counter-current drying exhibits a relatively slower initial drying rate followed by progressively increasing rates as the droplet approaches the hotter inlet air.

6.3 ANTISOLVENT VAPOR PRECIPITATION SPRAY DRYING

6.3.1 What Is This Concept About?

This technique was unexpectedly encountered in the attempt to generate porous crystalline lactose particles by exposing the lactose droplets to convective ethanol vapor using the single droplet drying technique described earlier in this book. The initial hypothesis was that the ethanol vapor absorbing into the droplet surface will cause the crystallization of the lactose on the surface of the droplet (owing to the insolubility of lactose in ethanolic conditions) forming a surface crust composed of ultrafine crystals. Unexpectedly, instead of achieving a crystalline crust, the final lactose particles resembled ultrafine spherical particles (Mansouri *et al.*, 2012). Further experiments that will be discussed later on revealed that spherical amorphous lactose or ultrafine crystalline lactose can be produced from a single droplet. Although not reported, the granular particles composed of these ultrafine particles are very loosely compacted, analogous to behaving like individual particles. This finding suggests that the spray drying process can potentially be used as a large-scale micro-precipitator, with each atomized droplet functioning like a precipitator (Figure 6.3.1).

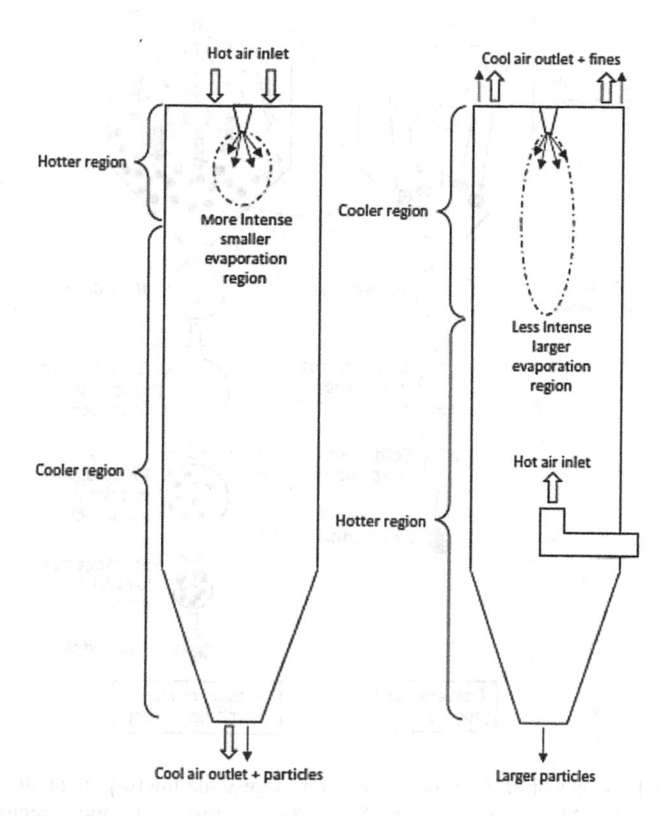

FIGURE 6.2.2 Differences between counter-current and co-current spray drying. (From Woo, M.W., Lee, M.G., Shakiba, S., Mansouri, S., Controlling in situ crystallization of pharmaceutical particles within the spray dryer. *Expert Opin. Drug Deliv.* 14, 1315–1324, 2017. With permission.)

One main advantage of this technique would then be the ability to control the maximum length scale of precipitation without the need for ultrafine geometries such as micro-channels or involving the use of membranes to produce ultrafine emulsions. On that note, one may also be thinking of spray drying such fine particulates directly. This would require very fine atomization to be used. The difficulty in generating such fine atomization can be deduced via the equation below.

$$R_{\text{initial}} = \sqrt[3]{\frac{\rho_{\text{final}} R^3_{\text{final}} X_{\text{final}}}{\rho_{\text{initial}} X_{\text{initial}}}} \tag{6.3.1}$$

If we consider a typical spray drying feed concentration of 20% wt, in order to produce a particle in the range of ~1 micron, the initial atomized droplets will have to be in the range of ~2 microns. It is possible for spray dryers to be fitted with an atomizer capable of producing droplets within this range of initial droplet sizes. However, such fine atomization is typically associated with low throughputs found in laboratory-scale spray dryers. Commercial high throughput spray dryers typically

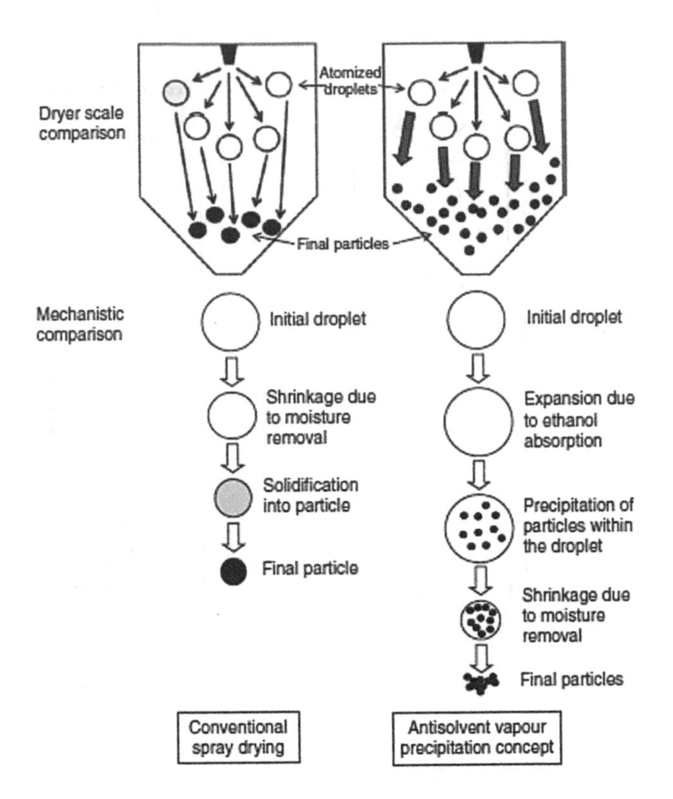

FIGURE 6.3.1 Concept of the spray dryer as a large-scale micro-precipitator technology. (From Woo, M.W., Mansouri, S., Chen, X.D., 2014. Antisolvent vapor precipitation: The future of pulmonary drug delivery particle production? *Expert Opin. Drug Del.* 11, 307–311, 2014. With permission.)

produce droplets in the range of tens to hundreds of microns. An alternative (to utilizing a high throughput commercial nozzle) would be to use a significantly higher concentration, but this may be limited by the allowable product formulation. In order to circumvent this limitation, the antisolvent vapor precipitation drying technique may potentially alleviate this problem by allowing ultrafine droplets to be formed using relatively large sprayed droplets and economically realistic initial feed concentrations (Woo *et al.*, 2014). In addition, large loosely agglomerated ultrafine particles will also make the handling and the flowability of the powder more easily manageable.

6.3.2 Unique Early Findings and Applications

The bulk of the reports so far on this new technique were obtained with single droplet drying experiments. Different types of materials were evaluated and, capitalizing on the various solubilities of the materials in the ethanolic environment, composite materials could also be produced. Some of these ultrafine particles are shown in Figure 6.3.2. It can be seen that, particularly from the illustration on reported

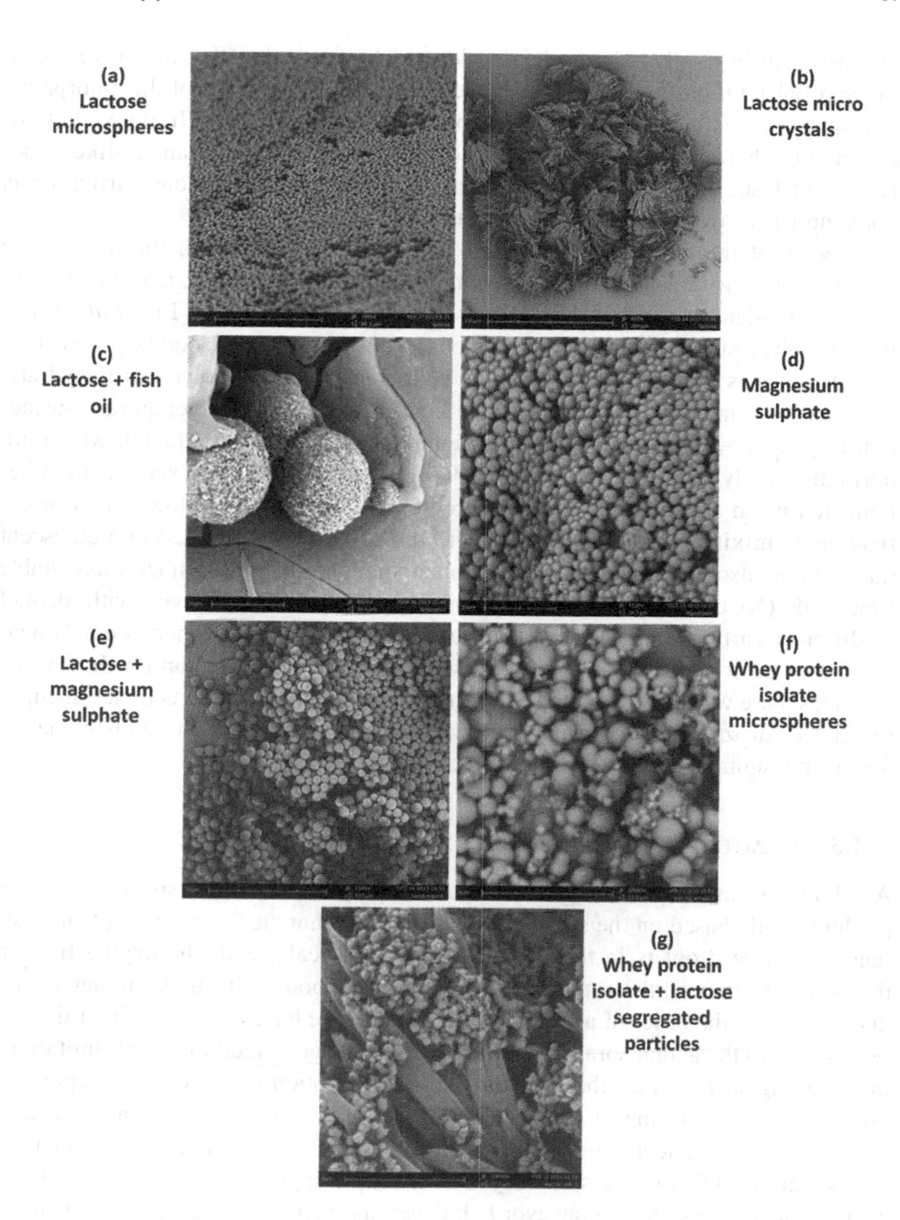

FIGURE 6.3.2 Ultrafine or composite precipitated particles. (From Woo, M.W., Mansouri, S., Chen, X.D., Antisolvent vapor precipitation: The future of pulmonary drug delivery particle production? *Expert Opin. Drug Del.* 11, 307–311, 2014. With permission.)

lactose experiments, crystalline or amorphous ultrafine particles can be produced. The key to controlling the antisolvent vapor precipitation process relied on controlling the absorption of the ethanol into the droplet; this has been undertaken thus far by adjusting the concentration of ethanol vapor in the convective airstream contacting the droplets. At the core of the process, the vapor concentration denotes the rate of mass transfer of ethanol into the droplet which affects the ethanol concentration in the droplet. Of particular interest, however, was the occurrence of the amorphous ultrafine particles in the intermediate vapor concentration region. In the vicinity of that region, there was also an incidence of the precipitation of filament-like structures, which suggested that the occurrence of the ultrafine amorphous particles was not simply just "oversaturation" precipitation of lactose.

In view of the fact that lactose is a "crystallizable" material, the next set of reported investigations then utilized a non-crystallization material, maltodextrin, to further investigate this potentially new precipitation phenomenon (Tan *et al.*, 2015). Fortuitously, a similar phenomenon was observed, which manifested as porous precipitated porous particles. This led to the confirmation that dissolved carbohydrate, e.g., lactose or maltodextrin, can actually behave like a transient temporary surfactant to generate a spontaneous unstable emulsion of water and ethanol, which are normally highly miscible. Herrington (1934) also accidentally observed this phenomenon when crystallizing lactose with ethanol. We further followed this investigation by mixing ethanol and water (with dissolved carbohydrates) in a quiescent manner and also observed this unusual phenomenon, albeit at a much more stable time scale (Xiao *et al.*, 2018). This phenomenon was observed even with alcohol of different carbon lengths, which further augmented the spontaneous emulsification behavior. It is noteworthy not to mistake this new phenomenon for the "ouzo" effect (as there was no oil component in the experimental evaluations undertaken) or the molecular-scale ethanol–water cage structure, as the emulsification phenomenon occurs at a significantly larger (micron) scale.

6.3.3 Challenges and Latest Developments

As alluded to in the preceding sections, the reports and findings so far have been predominantly based on the single droplet drying technique. The technique utilizes relatively large droplets in the range of microlitre scale, with the drying time in the order of minutes; the actual sprayed droplet is about 100 times smaller with a drying time in the order of a fraction of a second. We have yet to upscale this new technology to the actual spray drying scale, mainly due to technological limitations in designing such a dryer that will enable the absorption of the ethanol vapor, yet provide sufficient drying of the droplet. We have, however, experimentally evaluated whether the same unique precipitation phenomenon is possible at droplet length scales similar to that of the spray dryer. In a simple approach, we resorted to spray drying fine droplets onto a conveyor belt dryer and then allowing the droplets to be conveyed into an ethanol-rich section followed by a hot air-drying section (Chew, 2015). The advantage of this experimental technique was that the period of absorption and dehydration could be explicitly controlled. A similar unique precipitation phenomenon was observed. Therefore, this is evidence to suggest that the observations so far are applicable even at the real sprayed droplet scale.

Designing the spray dryer to achieve these requirements may require further groundwork to be undertaken. Firstly, from the reports so far, we have only managed to show how the ambient ethanol vapor concentration affects the final precipitated morphology of the dissolved solids in the droplet. In essence, the ambient vapor concentration affects the mass and heat transfer and it is the concentration of ethanol within the droplet (or the rate at which it changes) that actually affects precipitation behavior. There is still no report so far detailing how the droplet ethanol concentration or the rate at which it changes affects precipitation behavior, although we have some qualitative ideas about this relationship. A better understanding of this will be critical to determine the design criteria required for the antisolvent vapor precipitation spray dryer.

In any case, even if this understanding is available, there are still some gaps in the modeling of simultaneous ethanol–water vapor mass transfer. There are existing reports focusing on the modeling of the simultaneous evaporation of ethanol and water vapor from droplets. A key phenomenon in the antisolvent vapor precipitation process is the simultaneous ethanol absorption/water vapor evaporation in the initial duration of the droplet heat and as transfer. We have evaluated and found that the modeling of this process as a typical lump model is not trivial and that the simplistic approach of "switching" the direction of the driving force for the individual vapor was not sufficient to capture the phenomenon involved (Tan, 2015). Work is currently underway to evaluate suitable modeling approaches in this area. Once both areas highlighted above are more deeply investigated, only then can the design of spray dryers for antisolvent vapor precipitation be effectively undertaken.

On a separate note, we have also tried applying the same technique on hard to dissolve or hydrophobic-based material. For such a material, ethanol may be used as the solvent for the material, and steam (water) as the antisolvent (Mansouri *et al.*, 2014), in view of the fact that the dissolved material in the droplet has low solubility to water. Experimental trials have shown that this technique can effectively produce ultrafine drug materials.

References

Adhikari, B., Howes, T., Bhandari, B.R., Truong, V. 2004. Effect of addition of maltodextrin on drying kinetics and stickiness of sugar and acid-rich foods during convective drying: Experiments and modelling. *J. Food Eng.* 62, 53–68.

Afshar, S., Jubaer, H., Chen, B., Xiao, J., Chen, X.D., Woo, M.W. 2018. Computational fluid dynamics simulation of spray dryers: transient or steady state simulation? *Proceedings of the 21st International Drying Symposium*, September 11–14, 2018, Valencia, Spain.

Ali, M., Mahmud, T., Heggs, P.J., Ghadiri, M., Bayly, A., Ahmadian, H., Martin de Juan, L. 2017. CFD modeling of a pilot-scale countercurrent spray drying tower for the manufacture of detergent powder. *Drying Technol.* 35, 281–299.

Ali, M., Mahmud, T., Heggs, P.J., Ghadiri, M., Djurdjevic, D., Ahmadian, H., Martin de Juan, L., Amador, C., Bayly, A. 2014. A one-dimensional plug-flow model of a countercurrent spray drying tower. *Chem. Eng. Res. Des.* 92, 826–841.

Anandharamakrishnan, C., Gimbun, J., Stapley, A.G.F., Rielly, C.D. 2010. A study of particle histories during spray drying using computational fluid dynamic simulations. *Drying Technol.* 28, 566–576.

Blei, S., Sommerfeld, M. 2007. *CFD in Drying Technology – Spray-Dryer Simulation, Modern Drying Technology.* Wiley-VCH Verlag, Darmstadt,. 155–208.

Bowes, P.C. 1984. *Self-heating: Evaluating and Controlling the Hazards.* Elsevier, Amsterdam.

Buckton, G., Chidavaenzi, O.C., Fariba, K. 2002. The effect of spray-drying feed temperature and subsequent crystallization conditions on the physical form of lactose. *AAPS PharmSciTech* 3, 1–6.

Cal, K., Sollohub, K. 2010. Spray drying technique. I: hardware and process parameters. *J. Pharm. Sci.* 99(2), 575–586.

Camire, M.E., Dougherty, M.P. 2003. Raisins dietary fiber composition and in vitro bile acid binding. *J. Agric. Food Chem.* 51, 834–837.

Chen, G., Maier, D.E., Campanella, O.H., Takhar, P.S. 2009. Modeling of moisrure diffusivities for components of yellow-dent corn kernels. *J. Cereal Sci.* 50, 82–90.

Chen, X.D. 1992. On the mathematical modeling of the transient spontaneous heating in a moist coal stockpile. *Combust. Flame* 90, 114–120.

Chen, X.D. 1994. The effect of drying heat and moisture content on the maximum temperature rise during self-ignition of a moist coal pile. *Coal Prep.* 14, 223–236.

Chen, X.D. 1998. The fundamentals of self-ignition of water containing combustible materials. *Chem. Eng. Process.* 37, 367–378.

Chen, X.D. 2007. Simultaneous heat and mass transfer. In *Handbook of Food and Bioprocess Modeling Techniques*, edited by Sablani, S.S., Rahman, M.S., Datta, A.K., Mujumdar, A.S. CRC Press, Boca Raton, Chapter 6.

Chen, X.D. 2008. The basics of a reaction engineering approach to modeling air-drying of small droplets or thin layer materials. *Drying Technol.* 26(6), 627–639.

Chen, X.D., Chen, N. 1997. Preliminary introduction to a unified approach to modelling drying and equilibrium isotherms of moist porous solids. *Chemeca 1997, Rotorua, New Zealand* (on CD- ROM).

Chen, X.D., Lin, S.X.Q. 2005. Air drying of milk droplet under constant and time-dependent conditions. *AIChE J.* 51(6), 1790–1799.

Chen, X.D., Mujumdar, A.S. (Eds.) 2008. *Drying Technologies in Food Processing.* Blackwell Publishing, UK, Oxford.

Chen, X.D., Peng, X.F. 2005. Modified Biot number in the context of air drying of small moist porous objects. *Drying Technol.* 23, 83–103.

Chen, X.D., Pirini, W. 2004. The reaction engineering modelling approach to drying of thin layer of silica gel particles. In *Topics in Heat and Mass Transfer*, edited by Chen, G., Devahastin, S., Thorat, B.N. Vindhya Press, Mumbai, India, pp.131–140.

Chen, X.D., Stott, J.B. 1992. Calorimetric study of the heat of drying of a sub-bituminous coal. *J. Fire Sci.* 10, 352–361.

Chen, X.D., Xie, G.Z. 1997. Fingerprints of the drying of particulate or thin layer food materials established using a simple reaction engineering model. *Trans. Inst. Chem. Eng. Part C: Food Bio-Prod. Process.* 75(C), 213–222.

Chen, X.D., Pirini, W., Ozilgen, M. 2001. The reaction engineering approach to modeling drying of thin layer of pulped kiwifruit flesh under conditions of small Biot numbers. *Chem. Eng. Process.* 40, 311–320.

Chen, X.D., Sidhu, H., Nelson, M. 2011. Theoretical probing of the phenomenon of the formation of the outermost surface layer of a multi-component particle, and the surface chemical composition after the rapid removal of water in spray drying. *Chem. Eng. Sci.* 66, 6375–6384.

Chew, S. 2015. Microsphere formation in droplets using antisolvent vapour precipitation technique. Master's thesis, Monash University, Melbourne, Australia.

Chidavaenzi, O.C., Buckton, G., Koosha, F. 1998. The impact of feed temperature on the polymorphic content of spray dried lactose. *J. Pharm. Pharmacol.* 50, 184.

Chong, L.V., Chen, X.D. 1999. A mathematical model of the self-heating of spray-dried food powders containing fat, protein, sugar and moisture. *Chem. Eng. Sci.* 54(19), 4165–4178.

Das, D., Langrish, T.A.G. 2012. Combined crystallization and drying in a pilot-scale spray dryer. *Drying Technol.* 30, 998–1007.

Das, D., Husni, H.A., Langrish, T.A.G. 2010. The effects of operating conditions on lactose crystallization in a pilot-scale spray dryer. *J. Food Eng.* 100, 551–556.

De Vries, D.A. 1958. Simultaneous transfer of heat and moisture transfer in porous media. *Trans. Am. Geophys. Union* 39(5), 909–916.

Donovan, J.L., Meyer, A.S., Waterhouse, A.L. 1998. Phenolic composition and antioxidant activity of prunes and prune juice (Prunus domestica). *J. Agric. Food Chem.* 46(4), 1247–1252.

Ebrahimi, A., Langrish, T. 2015. Spray drying and crystallization of lactose with humid air in a straight-through system. *Drying Technol.* 33, 808–816.

Esfahani, S., Mansouri, S., Selomulya, C., Woo, M.W. 2018. The role of intermediate stage of drying on particle in-situ crystallization in spray dryers. *Powder Technol.* 323, 357–366.

Fäldt, P., Bergenståhl, B. 1994. The surface composition of spray-dried protein-lactose powders. *Colloids Surfaces A* 90, 183–190.

Fäldt, P., Bergenståhl, B. 1996a. Spray-dried whey protein/lactose/soybean oil emulsions. 1. Surface composition and particle structure. *Food Hydrocolloid* 10, 421–429.

Fäldt, P., Bergenståhl, B. 1996b. Spray-dried whey protein/lactose/soybean oil emulsions. 2. Redispersability, wettability and particle structure. *Food Hydrocolloid* 10, 431–439.

Feungfoo, M., Devahastin, S., Niamnuy, C., Soponronnarit, S. 2018. Preliminary study of superheated steam spray drying: A case study with maltodextrin. *Proceedings of the 21st International Drying Symposium,* September 11–14, 2018, Valencia, Spain, pp. 1147–1154.

Filkova, I., Huang, L.X., Mujumdar, A.S. 2015. Industrial spray drying systems. In *Handbook of Industrial Drying*, edited by Mujumdar, A.S. CRC Press, Boca Raton.

Fletcher, D., Langrish, T. 2009. Scale-adaptive simulation (SAS) modelling of a pilot-scale spray dryer. *Chem. Eng. Res. Des.* 87, 1371–1378.

Fletcher, D.F., Guo, B., Harvie, D.J.E., Langrish, T.A.G., Nijdam, J.J., Williams, J. 2006. What is important in the simulation of spray dryer performance and how do current CFD models perform? *Appl. Math. Model.* 30, 1281–1292.

Fogler, H.S. 1992. *Elements of Chemical Reaction Engineering*, 2nd edition. Prentice-Hall International, New Jersey, pp. 61–72.

Fortes, M., Okos, R. 1980. Drying theories: Their bases and limitations applied to food and grain. *Adv. Drying* 1, 119–154.

Francia, V., Martin, L., Bayly, A.E., Simmons, M.J.H. 2015. The role of wall deposition and re-entrainment in swirl spray dryers. *AIChE J.* 61, 1804–1821.

Frank, X., Perre, P. 2010. The potential of meshless methods to address physical and mechanical phenomena involved during drying at pore level. *Drying Technol.* 28, 932–943.

Frydman, A., Vasseur, J., Moureh, J., Sionneau, M., Tharrault, P. 1998. Comparison of superheated steam and air operated spray dryers using computational fluid dynamics. *Drying Technol.* 16, 1305–1338.

Fu, N., Chen, X.D. 2020. Chapter 14: Droplet drying. In *Modern Drying Technology*, 3rd ed., edited by Liu, X.-D., Li, Z. Chemical Industry Press, Beijing, China.

Fu, N., Huang, S., Xiao, J., Chen, X.D. 2018. Chapter Six: Producing powders containing active dry probiotics with the aid of spray drying, In *Advances in Food and Nutrition Research*, edited by Toldrá, F. Academic Press, pp. 211–262.

Fu, N., Woo, M.W., Chen, X.D. 2012. Single droplet drying technique to study drying kinetics measurement and particle functionality: a review. *Drying Technol.* 30(15), 1771–1785.

Fu, N., Woo, M.W., Lin, X.Q., Zhou, Z., Chen, X.D. 2011. Reaction engineering approach (REA) to model the drying kinetics of droplets with different initial sizes – experiments and analyses. *Chem. Eng. Sci.* 66, 1738–1747.

Gabites, J.R., Abrahamson, J., Winchester, J.A. 2010. Air flow patterns in an industrial milk powder spray dryer. *Chem. Eng. Res. Des.* 88, 899–910.

George, O.A., Chen X.D., Xiao, J., Woo, M., Che, L. 2015. An effective rate approach to modelling single-stage spray drying. *AIChE J.* 61(12), 4140–4151.

George, O.A., Xiao, J., Mercade-Prieto, R., Fu, N., Chen, X.D. 2019. Numerical probing of suspended lactose droplet drying experiment. *J. Food Eng.* 254, 51–63.

George, O.A., Xiao, J., Rodrigo, C.S., Mercade-Prieto, R., Sempere, J., Chen, X.D. 2017. Detailed numerical analysis of evaporation of a micrometer water droplet suspended on a glass filament. *Chem. Eng. Sci.* 165, 33–47.

Gianfrancesco, A., Turchiuli, C., Dumoulin, E., Patzer, S. 2009. Prediction of powder stickiness along spray drying process in relation to agglomeration. *Part. Sci. Technol.* 27, 415–427.

Gray, B.F. 1990. Analysis of chemical kinetic systems over the entire parameter space. III. A wet combustion system. *Proc. R. Soc. Lond. A* 429, 449–458.

Gray, B.F., Wake, G.C. 1990. The ignition of hygroscopic materials by water. *Combust. Flame* 79, 2–6.

Har, C.L., Fu, N., Chan, E.S., Tey, B.T., Chen, X.D. 2017. Unraveling the droplet drying characteristics of crystallization-prone mannitol – experiments and modeling. *AICHE J.* 63(6), 1839–1852.

Harvie, D.J.E., Langrish, T.A.G., Fletcher, D.F. 2002. A computational fluid dynamics study of a tall-form spray dryer. *Chem. Eng. Res. Des.* 80, 163–175.

Herrington, B.L. 1934. Some physico-chemical properties of lactose. I. The spontaneous crystallization of supersaturated solutions of lactose. *J. Dairy Sci.* 17(7), 501–518.

Hogan, S., O'Callaghan, D. 2013. Moisture sorption and stickiness behavior of hydrolysed whey protein/lactose powders. *Dairy Sci. Technol.* 93, 505–521.

Huang, L., Kumar, K., Mujumdar, A.S. 2004. Simulation of a spray dryer fitted with a rotary disk atomizer using a three-dimensional computational fluid dynamic model. *Drying Technol.* 22, 1489–1515.

Huang, L.X., Mujumdar, A.S., Chen, X.D. 2010. An overview of the recent advances in spray drying. *Dairy Sci. Technol.* 90(2–3), 211–224.

Hui, Y.H. 2006. *Handbook of Fruits and Fruit Processing*. Blackwell Publishing, Oxford, UK, 81.

Incropera, F.P., Dewitt, D.P. 1990. *Fundamentals of Heat and Mass Transfer*, 4th edition. Wiley, New York.

Islam, M.I.U., Langrish, T.A.G. 2010. An investigation into lactose crystallization under high temperature conditions during spray drying. *Food Res. Int.* 43, 46–56.

Islam, M.I.U., Langrish, T.A.G., Chiou, D. 2010. Particle crystallization during spray drying in humid air. *J. Food Eng.* 99, 55–62.

Jaskulski, M., Atuonwu, J.C., Tran, T.T.H., Stapley, A.G.F., Tsotsas, E. 2017. Predictive CFD modeling of whey protein denaturation in skim milk spray drying powder production. *Adv. Powder Technol.* 28, 3140–3147.

Jaskulski, M., Wawrzyniak, P., Zbiciński, I. 2015. CFD model of particle agglomeration in spray drying. *Drying Technol.* 33, 1971–1980.

Jin, Y., Chen, X.D. 2009a. A three-dimensional numerical study of the gas/particle interactions in an industrial-scale spray dryer for milk powder production. *Drying Technol.* 27, 1018–1027.

Jin, Y., Chen, X.D. 2009b. Numerical study of the spray drying process of different sized particles in an industrial-scale spray dryer. *Drying Technol.* 27, 371–381.

Jin, Y., Chen, X.D. 2010. A fundamental model of particle deposition incorporated in CFD simulations of an industrial milk spray dryer. *Drying Technol.* 28, 960–971.

Jin, Y., Chen, X.D. 2011. Entropy production during drying process of milk droplets in an industrial spray dryer. *Int. J. Therm. Sci.* 50, 615–625.

Jongsma, F., Innings, F., Olsson, M., Carlsson, F. 2013. Large eddy simulation of unsteady turbulent flow in a semi-industrial size spray dryer. *Dairy Sci. Technol.* 93, 373–386.

Jubaer, H., Afshar, S., Xiao, J., Chen, X.D., Selomulya, C., Woo, M.W. 2017. On the importance of droplet shrinkage in CFD-modelling of spray drying. *Drying Technol.* Doi: 10.1080/07373937.2017.1349791.

Jubaer, H., Dai, R., Ruslim, E., Shahnaz, M., Shan, Z., Woo, M.W. 2019a. New perspectives on capturing particle agglomerates in CFD modelling of spray dryers. *Drying Technol.* 38 (5–6), 685–694.

Jubaer, H., Xiao, J., Chen, X.D., Selomulya, C., Woo, M.W. 2019b. Identifications of regions in a spray dryer susceptible to forced agglomeration by CFD simulations. *Powder Technol.* 346, 23–37.

Jubaer, H., Afshar, S., Xiao, J., Chen, X.D., Selomulya, C., Woo, M.W. 2019c. On the effect of turbulence models on CFD simulations of a counter-current spray drying process. *Chemical Engineering Research and Design.* 141, 592–607.

Kar, S. and Chen, X.D. 2009. The impact of various drying kinetics models on the prediction of sample temperature-time and moisture content-time profiles during moisture removal from stratum corneum. *Chem. Eng. Res. Des.* 87, 739–755.

Keey, R.B. 1992. *Drying of Particulate and Loose Materials*. Hemisphere, New York.

Kentish, S., Davidson, M., Hassan, H., Bloore, C. 2005. Milk skin formation during drying. *Chem. Eng. Sci.* 60, 635–646.

Kieviet, F.G., Van Raaij, J., De Moor, P.P.E.A., Kerkhof, P.J.A.M. 1997. Measurement and modelling of the air flow pattern in a pilot-scale spray dryer. *Chem. Eng. Res. Des.* 75, 321–328.

Kim, E.H.-J., Chen, X.D., Pearce, D. 2002. Surface characterization of four industrial spray-dried dairy powders in relation to chemical composition, structure and wetting property. *Colloid. Surface. B.* 26, 197–212.

Kim, E.H.J., Chen, X.D., Pearce, D. 2003. On the mechanisms of surface formation and the surface compositions of industrial milk powders. *Drying Technol.* 21, 265–278.

Kota, K., Langrish, T.A.G. 2007. Prediction of deposition patterns in a pilot-scale spray dryer using computational fluid dynamics (CFD) simulations. *Chem. Prod. Proc. Model.* 2, Article 26.

Langrish, T.A.G., Kockel, T.K. 2001. The assessment of a characteristic drying curve for milk powder for use in computational fluid dynamics modelling. *Chem. Eng. J.* 84, 69–74.

Langrish, T.A.G., Williams, J., Fletcher, D.F. 2004. Simulation of the effects of inlet swirl on gas flow patterns in a pilot-scale spray dryer. *Chem. Eng. Res. Des.* 82, 821–833.

Lin, S.X.Q., Chen, X.D. 2005. Prediction of air drying of milk droplet under relatively high humidity using the reaction engineering approach. *Drying Technol.* 23(7), 1395–1406.

Lin, S.X.Q., Chen, X.D. 2006. A model for drying of an aqueous lactose droplet using the reactionengineering approach. *Drying Technol.* 24, 1329–1334.

Lin, S.X.Q., Chen, X.D. 2007. The reaction engineering approach to modeling the cream and whey protein concentrate droplet drying. *Chem. Eng. Process.* 46, 437–443.

Lin, R.H., Liu, W.J., Woo, W.M., Chen, X.D., Selomulya, C. 2015. On the formation of "coral-like" spherical alpha-glycine crystalline particles. *Powder Technol.* 279, 310–316.

Littringer, E.M., Mescher, A., Schroettner, H., Achelis, L., Walzel, P., Urbanetz, N.A. 2012. Spray dried mannitol carrier particles with tailored surface properties – the influence of carrier surface roughness and shape. *Eur. J. Pharm. Biopharm.* 82, 194–204.

Littringer, E.M., Paus, R., Mescher, A., Schrottner, H., Walzel, P., Urbanetz, N.A. 2013. The morphology of spray dried mannitol particles – the vital importance of droplet size. *Powder Technol.* 239, 162–174.

Luikov, A.V. 1975. *Heat and Mass Transfer in Capillary-Porous Bodies*. Pergamon Press, Oxford.

Luikov, A.V. 1986. *Drying Theory*. Energia, Moscow.

Lum, A. 2019. Superheated steam in spray drying for particle functionality engineering. PhD thesis, Monash University, Melbourne, Australia.

Lum, A., Cardamon, N., Beliavski, R., Mansouri, S., Hapgood, K., Woo, M.W. 2018. Unusual drying behaviour of droplets containing organic and inorganic solutes in superheated steam. *J. Food Eng.* 244, 64–72.

Lum, A., Cardamone, N., Beliavski, R., Mansouri, S., Hapgood, K., Woo, M.W. 2019. The role of steam as a medium for droplet crystallization. *Ind. Eng. Chem. Res.* (accepted: 23 April 2019).

Lum, A., Mansouri, S., Hapgood, K., Woo, M.W. 2017. Single droplet drying of milk in air and superheated steam: Particle formation and wettability. *Drying Technol.* 36, 1802–1813.

Maas, S.G., Schaldach, G., Littringer, E.M., Mescher, A., Griesser, U.J., Braun, D.E., Walzel, P.E., Urbanetz, N.A. 2011. The impact of spray drying outlet temperature on the particle morphology of mannitol. *Powder Technol.* 213, 27–35.

Malafronte, L., Ahrne, L., Innings, F., Jongsma, A., Rasmuson, A. 2015. Prediction of regions of coalescence and agglomeration along a spray dryer-application to skim milk powder. *Chem. Eng. Res. Des.* 104, 703–712.

Mansouri, S., Chin, G.Q., Ching, T.W., Woo, M.W., Fu, N., Chen, X.D. 2013. Precipitating smooth amorphous or pollen structured lactose microparticles. *Chem. Eng. J.* 226, 312–318.

Mansouri, S., Fu, N., Woo, M.W., Chen, X.D. 2012. Uniform amorphous lactose microspheres formed simultaneous convective and dehydration antisolvent precipitation under atmospheric conditions. *Langmuir* 28, 13772–13776.

Mansouri, S., Hena, V.S., Woo, M.W. 2015. Narrow tube spray drying. *Drying Technol.* 34(9), 1043–1051.

Mansouri, S., Kralj, T.P., Morton, D., Chen, X.D., Woo, M.W. 2014. Squeezing out ultrafine hydrophobic and poor water-soluble drug particles with water vapour. *Adv. Powder Technol.* 25, 1190–1194.

Masters, K. 1979. *Spray Drying Handbook*. George Godwin, London.

Matsumoto, M., Yasuoka, K., Kataoka, Y. 1995. Molecular simulation of evaporation and condensation. *Fluid Phase Equilibr.* 104(1995), 431–439.

Mezhericher, M., Levy, A., Borde, I. 2009. Modeling of droplet drying in spray chambers using 2D and 3D computational fluid dynamics. *Drying Technol.* 27, 359–370.

Mezhericher, M., Levy, A., Borde, I. 2012. Probabilistic hard-sphere model of binary particle–particle interactions in multiphase flow of spray dryers. *Int. J. Multiphase Flow* 43, 22–38.

Mezhericher, M., Levy, A., Borde, I. 2015. Multi-scale multiphase modeling of transport phenomena in spray-drying processes. *Drying Technol.* 33, 2–23.

Nijdam, J.J., Langrish, T.A.G. 2006. The effect of surface composition on the functional properties of milk powders. *J. Food Eng.* 77, 919–925.

Oakley, D.E., Bahu, R.E. 1993. Computer modelling of spray dryers. *Comput. Chem. Eng.* 17, S493–S498.

Ortega-Rivas, E., Juliano, P., Yan, H. 2006. *Food Powders: Physical Properties, Processing, and Functionality.* Springer Science & Business Mediax, New York.

Ozmen, L., Langrish, T.A.G. 2003. A study of the limitations to spray dryer outlet performance. *Drying Technol.* 21, 895–917.

Patel, K., Chen, X.D., Jeantet, R., Schuck, P. 2010. One-dimensional simulation of co-current, dairy spray drying systems–pros and cons. *Dairy Sci. Technol.* 90, 181–210.

Patel, K.C. 2004. A novel concept of spray drying – modelling and design. ME thesis, Chemical Engineering, University of Auckland, New Zealand (supervised by X.D. Chen).

Patel, K.C. 2008. Production of uniform particles via single stream drying and new applications of the reaction engineering approach. PhD thesis, Chemical Engineering, Monash University, Melbourne, Australia (supervised by X.D. Chen)

Patel, K.C., Chen, X.D., Lin, S.X.Q., Adhikari, B. 2009a. A composite reaction engineering approach to drying of aqueous droplets containing sucrose, maltodextrin(DE6), and their mixtures. *AIChE J.* 55, 217–231.

Patel, R., Patel, M., Suthar, A. 2009b. Spray drying technology: An overview. *Ind. J. Sci. Technol.* 2(10), 44–47.

Perre, P. 2011. A review of modern computational and experimental tools relevant to the field of drying. *Drying Technol.* 29, 1529–1541.

Petersen, L.N., Poulsen, N.K., Niemann, H.H., Utzen, C., Jorgensen, J.B. 2017. An experimentally validated simulation model for a four-stage spray dryer. *J. Process Control* 57, 50–65.

Philip, J.R., De Vries, D.A. 1957. Moisture movement in porous materials under temperature gradients. *Trans. Am. Geophys. Union* 38(5), 222–232, 594.

Putranto, A., Chen, X.D., Webley, P. 2009. Infrared and convective drying of thin layer of polyvinyl alcohol (PVA)/glycerol/water mixture – the reaction engineering approach (REA). *Chem. Eng. Process.* 49, 348–357.

Putranto, A., Chen, X.D., Webley, P.A. 2010. Infrared and convective drying of thin layer of polyvinyl alcohol (PVA)/glycerol/water mixture - The reaction engineering approach (REA). *Chem. Eng. Process.* 49, 348–357.

Putranto, A., Chen, X.D., Xiao, Z.Y., Davastin, S., Webley, P.A. 2011. Application of the REA (reaction engineering approach) for modelling intermittent drying under time-varying humidity and temperature. *Chem. Eng. Sci.* 66(10), 2149–2156.

Quek, C.X.L. 2011. Operation of falling film evaporator involving viscous liquid. PhD thesis, Monash University, Melbourne, Australia.

Razmi, R., Yu, W., Young, B., Woo, M.W. 2019. What is important in the design of counter current spray drying towers? *Proceedings of CHEMECA,* September 29–October 2, 2019, Sydney, Australia.

Rogers, S. 2011. Developing and utilizing a mini food powder production facility to produce industrially relevant particles for functionality testing. PhD thesis, Chemical Engineering, Monash University, Melbourne, Australia (supervised by X.D. Chen).

Rogers, S., Wu, W.D., Lin, S.X.Q., Chen, X.D. 2012. Particle shrinkage and morphology of milk powder made with a monodisperse spray dryer. *Biochem. Eng. J.* 62, 92–100.

Rostami, A., Murthy, J., Hajaligol, M. 2003. Modeling of a smoldering cigarette. *J. Anal. Appl. Pyrol.* 66, 281–301.

Sander, A, Penović, T, Šipušić, J. 2011. Crystallization of β-glycine by spray drying. *Cryst. Res. Technol.* 46, 145–152.

Santos, D., Maurício, A.C., Sencadas, V., Santos, J.D., Fernandes, M.H., Gomes, P.S. 2018. Spray drying: An overview. In *Biomaterials – Physics and Chemistry – New Edition.* Doi: 10.5772/intechopen.72247.

Schmitz-Schug, I., Kulozik, U., Foerst, P. 2016. Modeling spray drying of dairy products – impact of drying kinetics, reaction kinetics and spray drying conditions on lysine loss. *Chem. Eng. Sci.* 141, 315–329.

Schuck, P., Dolivet, A., Mejean, S., Zhu, P., Blanchard, E., Jeantet, R. 2009. Drying by desorption: A tool to determine spray drying parameters. *J. Food Eng.* 94, 199–204.

Schuck, P., Roignant, M., Brulé, G., Davenel, A., Famelart, M.H., Maubois, J.L. 1998. Simulation of water transfer in spray drying. *Drying Technol.* 16(7), 1371–1393.

Shakiba, S., Mansouri, S., Selomulya, C., Woo, M.W. 2019. Time scale based analysis of in-situ crystal formation in droplet undergoing rapid dehydration. *Int. J. Pharm.* 50, 47–56.

Shang, L.C., Chen, X.D., Xiao, J. 2019. Coarse-grained simulation of surface morphology formation for spray dried particles. *CIESC J.* 70(6), 2153–2163.

Shen, D., Bulow, M., Siperstein, F., Engelhard, M., Myers, A.L. 2000. Comparison of experimental techniques for measuring isosteric heat of adsorption. *Adsorption* 6, 275–286.

Sommerfeld, M., Stubing, S. 2017. A novel Lagrangian agglomerate structure model. *Powder Technol.* 319, 34–52.

Straatsma, J., van Houwelingen, G., Meulman, A.P., Steenbergen, A.E. 1991. Dryspec2: A computer model of a two stage dryer. *J. Soc. Dairy Technol.* 44, 107–111.

Straatsma, J., Van Houwelingen, G., Steenbergen, A.E., De Jong, P. 1999. Spray drying of food products: Simulation model. *J. Food Eng.* 42, 67–72.

Sulsky, D., Chen, Z., Schreyer, H.L. 1994. A particle method for history-dependent materials. *Comput. Methods Appl. Mech. Eng.* 118, 179–196.

Sulsky, D., Zhou, S.J., Schreyer, H.L. 1995. Application of a particle-in-cell method to solid mechanics. *Comput. Phys. Commun.* 87, 236–252.

Takhar, P.S., Maier, D.E., Campanella, O.H., Chen, G. 2011. Hybrid mixture theory based moisture transport and stress development in corn kernels during drying: Validation and simulation results. *J. Food Eng.* 106, 275–282.

Tan, J.Y. 2015. Mechanism of the novel antisolvent vapour precipitation (AVP) process. Master's thesis, Monash University, Melbourne, Australia.

Tan, J.Y., Tang, V.M., Nguyen, J., Chew, S., Mansouri, S., Hapgood, K., Chen, X.D., Woo, M.W. 2015. Unveiling the mechanism of antisolvent vapour precipitation in producing ultrafine spherical particles. *Powder Technol.* 275, 152–160.

Tinker, L.F., Schneeman B.O., Davis, P.A., Gallaher, D.D., Waggoner, C.R. 1991. Consumption of prunes as a source of dietary fiber in men with mild hypercholesterolemia. *Am. J. Clin. Nutr.* 53, 1259–1265.

Truong, V., Bhandari, B., Howes, T. 2005. Optimization of co-current spray drying process of sugar-rich foods. Part 1 – moisture and glass transition temperature profile during drying. *J. Food Eng.* 71, 55–65.

Tsotsas, E., Mujumdar, A.S. 2007. Preface. In *Modern Drying Technology. Volume 1: Computational Tools at Different Scales*, edited by Tsotsas, E., Mujumdar, A.S. Wiley-VCH, Weinheim, Germany, pp. XV–XVIII.

Ullum, T., Sloth, J., Brask, A., Wahlberg, M. 2010. Predicting spray dryer deposits by CFD and an empirical drying model. *Drying Technol.* 28, 723–729.

USDA. 2007. USDA database for the flavonoid content of selected foods (Release 2.1 2007).

Vehring, R. 2008. Pharmaceutical particle engineering via spray drying. *Pharm. Res.* 25, 999–1022.

Vehring, R., Foss, W.R., Lechuga-Ballesteros, D. 2007. Particle formation in spray drying. *Aerosol Sci.* 38, 728–746.

Verdumen, R.E.M., Menn, P., Ritzert, J., Blei, S., Nhumaio, G.C.S., Sorenson, T.S., Gunsing, M., Straatsma, J., Verschueren, M., Sibeijn, M., Schulte, G., Fritsching, U., Bauckhage, K., Tropea, C., Sommerfeld, M., Watkins, A.P., Yule, A.J., Schonfeldt, H. 2004. Simulation of agglomeration in spray drying installations: The EDECAD project. *Drying Technol.* 22, 1403–1461.

Vinson, J.A., Zubic, L., Bose, P., Samman, N., Proch, J. 2005. Dried fruits: Excellent in vivo and in vitro antioxidants. *J. Am. Coll. Nutr.* 24, 44–50.

Wang, S., Langrish T., Adhikari, B. 2013. A multicomponent distributed parameter model for spray drying: Model development and validation with experiments. *Drying Technol.* 31(13–14), 1513–1524.

Wang, S., Langrish, T.A.G. 2009. A distributed parameter model for particles in the spray drying process. *Adv. Powder Technol.* 20, 220–226.

Wang, Z., Zhou, Y., Li, S., Liu, J. (Eds.) 2007. *Physical Chemistry*, Volume 2, 4th edition. Higher Education Press, Beijing, pp. 196–223 (in Chinese).

Wawrzyniak, P., Jaskulski, M., Zbiciński, I., Podyma, M. 2017. CFD modelling of moisture evaporation in an industrial dispersed system. *Adv. Powder Technol.* 28, 167–176.

Wawrzyniak, P., Podyma, M., Zbicinski, I., Bartczak, Z., Rabaeva, J. 2012. Modeling of air flow in an industrial countercurrent spray-drying tower. *Drying Technol.* 30, 217–224.

Wei, Y.C., Woo, M.W., Selomulya, C., Wu, W.D., Xiao, J., Chen, X.D. 2019. Numerical simulation of mono-disperse droplet spray dryer under the influence of nozzle motion. *Powder Technol.* 355, 93–105.

Whitaker, S. 1977. Simultaneous heat, mass and momentum transfer in porous media. A theory of drying. *Adv. Heat Transf.* 13, 119–203.

Whitaker, S. 1999. *The Method of Volume Averaging.* Kluwer Academic Publishers, Dordrecht.

Williamson, G., Carughi, A. 2010. Polyphenol content and health benefits of raisins. *Nutr. Res.* 30, 511–519.

Woo, M.W. 2016. *Computational Fluid Dynamic Simulation of Spray Dryers – An Engineer's Guide.* CRC Press, Boca Raton, FL.

Woo, M.W. 2019. Advances in production of food powders by spray drying. In *Advanced Drying Technologies for Food*, edited by Mujumdar, A.S., Xiao, H.W. Taylor & Francis, Boca Raton.

Woo, M.W., Daud, W.R.W., Mujumdar, A.S., Talib, M.Z.M., Wu, Z.H., Tasirin, S.M. 2008a. Comparative study of drying models for CFD simulations of spray dryers. *Chem. Eng. Res. Des.* 86, 1038–1048.

Woo, M.W., Daud, W.R.W., Mujumdar, A.S., Talib, M.Z.M., Wu, Z.H., Tasirin, S.M. 2009. Non-swirling steady and transient flow simulations in short-form spray dryers. *Chem. Prod. Process. Model.* 4, 20.

Woo, M.W., Daud, W.R.W., Mujumdar, A.S., Tasirin, S.M., Talib, M.Z.M. 2010. The role of rheology characteristics in amorphous food particle-wall collisions in spray drying. *Powder Technol.* 198, 251–257.

Woo, M.W., Daud, W.R.W., Mujumdar, A.S., Wu, Z.H., Talib, M.Z.M., Tasirin, S.M. 2008b. CFD evaluation of droplet drying models in a spray dryer fitted with a rotary atomizer. *Drying Technol.* 26, 1180–1198.

Woo, M.W., Fu, N., Moo, R., Chen, X.D. 2012a. Unveiling the mechanisms of in-situ crystallization in spray drying of sugars. *Ind. Eng. Chem. Res.* 51, 11791–11802.

Woo, M.W., Le, C.M., Daud, W.R.W., Mujumdar, A.S., Chen, X.D., Tasirin, S.M., Talib, M.Z.M. 2012b. High swirling transient flows in spray dryer and consequent effect on modeling of particle deposition. *Chem. Eng. Res. Des.* 90, 336–345.

Woo, M.W., Lee, M.G., Shakiba, S., Mansouri, S. 2017. Controlling in situ crystallization of pharmaceutical particles within the spray dryer. *Expert Opin. Drug Deliv.* 14, 1315–1324.

Woo, M.W., Mansouri, S., Chen, X.D. 2014. Antisolvent vapour precipitation: The future of pulmonary drug delivery particle production? *Expert Opin. Drug Del.* 11, 307–311.

Woo, M.W., Rogers, S., Lin, S.X.Q., Selomulya, C., Chen, X.D. 2011a. Numerical probing of a low velocity concurrent pilot scale spray drying tower for mono-disperse particle production – unusual characteristics and possible improvements. *Chem. Eng. Process.* 50, 417–427.

Woo, M.W., Rogers, S., Selomulya, C., Chen, X.D. 2011b. Particle drying and crystallization characteristics in a low velocity concurrent pilot scale spray drying tower. *Powder Technol.* 223, 39–45.

Wu, W.D. 2010. A novel micro-fluidic-jet-spray-dryer equipped with a micro-fluidic-aerosol-nozzle: equipment development and applications in making functional particles. PhD thesis, Chemical Engineering, Monash University, Melbourne, Australia (supervised by X.D. Chen).

Wu, W.D., Lin, S.X., Chen, X.D. 2011. Monodisperse droplet formation through a continuous jet break-up using glass nozzles operated with piezoelectric pulsation. *AIChE J.* 57(6), 1386–1392.

Wu, W.D., Liu, W., Gengenbach, T., Woo, M.W., Selomulya, C., Chen, X.D., Weeks, M. 2014. Towards spray drying of high solids dairy liquid: effects of feed solid content on particle structure and functionality. *J. Food Eng.* 123, 130–135.

Wu, X., Beecher, G.R., Holden, J.M., Haytowitz, D.B., Gebhardt, S.E., Prior, R.L. 2004. Lipophilic and hydrophilic antioxidant capacities of common foods in the United States. *J. Agric. Food Chem.* 52, 4026–4037.

Xiao, J., Chen, L., Wu, W.D., Chen, X.D. 2016. Multiscale modeling for nano-scale surface composition of spray-dried powders: the effect of initial droplet size. *Drying Technol.* 34(9), 1063–1072.

Xiao, J., Chen, X.D. 2014. Multiscale modelling for surface composition of spray-dried two-component powders. *AIChE J.* 60, 2416–2427.

Xiao, J., Li, Y., George, O.A., Li, Z.H., Yang, S.J., Woo, M.W., Wu, W.D., Chen, X.D. 2018a. Numerical investigation of droplet pre-dispersion in a monodisperse droplet spray dryer. *Particuology* 38, 44–60.

Xiao, J., Yang, S.J., George, O.A., Putranto, A., Wu, W.D., Chen, X.D. 2019. Numerical simulation of mono-disperse droplet spray dryer: Coupling distinctively different sized chambers. *Chem. Eng. Sci.* 200, 12–26.

Xiao, J., Zhang, H., Wu, W.D., Chen, X.D. 2015. An improved calculation procedure on surface composition of spray dried protein-sugar systems. *Drying Technol.* 33(7), 817–821.

Xiao, P.L., Mansouri, S., SuriyaHena, V., Bong, Y.K., Hapgood, K., Woo, M.W. 2018b. Spontaneous ethanol-water emulsification as a precursor for porous particle formation: understanding the role of dissolved carbohydrates. *Particuology* 44, 44–53.

Yang, S.F., Xiao, J., Woo, M.W., Chen, X.D. 2015. Three-dimensional numerical investigation of a mono-disperse droplet spray dryer: validation aspects and multi-physics exploration. *Drying Technol.* 33, 742–756.

Yang, S.J., Wei, Y.C., Woo, M.W., Wu, W.D., Chen, X.D., Xiao, J. 2018. Numerical simulation of mono-disperse droplet spray dryer under influence of swirling flow. *CIESC J.* 69(9), 3814–3824.

Zbicinski, I., Li, X. 2006. Conditions for accurate CFD modelling of spray-drying process. *Drying Technol.* 24, 1109–1114.

Zheng, X., Fu, N., Duan, M., Woo, M.W., Selomulya, C., Chen, X.D. 2015. The mechanisms of the protective effects of reconstituted skim milk during convective droplet drying of lactic acid bacteria. *Food Res. Int.* 76, 478–488.

Zhu, P., Méjean, S., Blanchard, E., Jeantet, R., Schuck, P. 2011a. Prediction of dry mass glass transition temperature and the spray drying behaviour of a concentrate using a desorption method. *J. Food Eng.* 105, 460–467.

Zhu, P., Patel, K., Lin, S., Méjean, S., Blanchard, E., Chen, X.D., Schuck, P., Jeantet, R. 2011b. Simulating industrial spray drying operations using a reaction engineering approach and a modified desorption method. *Drying Technol.* 29(04), 419–428.

Zhu, P., Méjean, S., Blanchard, E., Jeantet, R., Schuck, P. 2013. Prediction of drying of dairy products using a modified balance-based desorption method. *Dairy Sci. Technol.* 93(4), 347–355.

Zuo, J.Y., Paterson, A.H., Bronlund, J.E., Chatterjee, R. 2007. Using a particle-gun to measure initiation of stickiness of dairy powders. *Int. Dairy J.* 17, 268–273.

Index

Printed in the United States
by Baker & Taylor Publisher Services